Dealing with Darwin

MEDICINE, SCIENCE, AND RELIGION
IN HISTORICAL CONTEXT

Ronald L. Numbers, *Consulting Editor*

Dealing with Darwin

Place, Politics, and Rhetoric in Religious Engagements with Evolution

David N. Livingstone

THE GIFFORD LECTURES, 2014

JOHNS HOPKINS UNIVERSITY PRESS   BALTIMORE

© 2014 Johns Hopkins University Press
All rights reserved. Published 2014
Printed in the United States of America on acid-free paper
9 8 7 6 5 4 3 2 1

Johns Hopkins University Press
2715 North Charles Street
Baltimore, Maryland 21218-4363
www.press.jhu.edu

Library of Congress Cataloging-in-Publication Data

Livingstone, David N., 1953–
    Dealing with Darwin : place, politics, and rhetoric in religious engagements
with evolution / David N. Livingstone.
        pages cm. — (Medicine, science, and religion in historical context)
    Includes bibliographical references and index.
    ISBN-13: 978-1-4214-1326-6 (hardcover : alk. paper)
    ISBN-13: 978-1-4214-1327-3 (electronic)
    ISBN-10: 1-4214-1326-4 (hardcover : alk. paper)
    ISBN-10: 1-4214-1327-2 (electronic)
    1. Evolution—Religious aspects—Christianity.   2. Darwin, Charles,
    1809–1882.   I. Title.
    BT712.L585 1987
    231.7'652088285—dc23   2013028976

A catalog record for this book is available from the British Library.

*Special discounts are available for bulk purchases of this book. For more information,
please contact Special Sales at 410-516-6936 or specialsales@press.jhu.edu.*

Johns Hopkins University Press uses environmentally friendly book materials,
including recycled text paper that is composed of at least 30 percent post-
consumer waste, whenever possible.

For Frances

CONTENTS

This book has been more than a dozen years in the making. In a conversation one afternoon in March 1999 in Berkeley, California, my good friend Ron Numbers talked me into writing two books. One was on the history of the idea of humans before Adam. It eventually saw the light of day in 2008 under the title *Adam's Ancestors*. This is the other book. At that stage I had written a few articles on the ways in which Darwin's theory was encountered in different settings and was working on a short book about the spaces of scientific inquiry. Ron thought that it might be worth trying to gather these thoughts together into a fuller account of what I was beginning to call "the geographies of Darwinism and religion."

I am more grateful than I can say to Ron for the prompt. For when the invitation to deliver the 2014 Gifford Lectures came along, that seemed the ideal venue for trying out these ideas. In a speech space that has been occupied mostly by philosophers, historians, and theologians, the thought that bringing a geographical perspective to bear on the whole issue of the encounter between science and religion struck me as novel . . . if risky! Still, the project of locating debates over evolution in particular sites and situations seemed to me a useful way of going beyond stereotype and caricature in understanding how religious communities dealt with Darwin and of ascertaining the role played by what I call place, politics, and rhetoric in public encounters with one of the greatest scientific theories of our time. So I am enormously indebted to Dr. Philip Ziegler and the other members of the Gifford Lectures Committee at the University of Aberdeen for presenting me with such a wonderful opportunity to share my thoughts on this whole subject.

On my journey I have incurred many other debts too. First and foremost I am grateful to Ron Numbers, not only for the stimulus to write the book but also for his generosity in making available to me his collection of materials on the James Woodrow case. For assistance with other parts of the story, I am grateful to Kim Arnold, archivist of the Presbyterian Church in Canada, Toronto, who willingly responded to various queries on Knox College; to Magomat Philemon of the Chancellor Oppenheimer Library, University of Cape Town, who provided some South

African material; to Wesley Bonnar and the staff at Sentry Hill, for their assistance with my research on Samuel B. G. McKinney; to Stephen Gregory of the Union College Library for facilitating various aspects of my research on the Tyndall episode. Patrick Harries of the University of Basel provided me with a number of fertile leads, and John Stenhouse, University of Otago, willingly made some of his unpublished work available to me. The late Ernan McMullin, whom I had the privilege of meeting frequently after he returned to Ireland, read and commented astutely on parts of the manuscript. I still enormously miss his wise counsel and penetrating wit. My good friend Mark Noll, now at Notre Dame, has patiently listened to the substance of this story on many occasions and, with the instincts of the historian, has probed me—hopefully to good effect—on my geographical predilections. John Hedley Brooke and Janet Browne read the entire manuscript and offered extremely valuable insights. I am very much in their debt. In Belfast I have been fortunate to be surrounded by a group of scholars with different disciplinary interests who combine in full measure critical insight tempered with enthusiastic encouragement: John Brewer, Diarmid Finnegan, Frank Gourley, Andrew Holmes, Nuala Johnson, Philip Orr, and Stephen Williams, and when he has been visiting Queen's University from UCLA, John Agnew. Their willingness to read what I write and listen to my enthusiasms means more than I can say.

But as ever, my greatest debt is to Frances, who has borne with great patience and good humor the challenges of living with a husband who spends too much time in the nineteenth century and not enough doing chores in the twenty-first.

Dealing with Darwin

CHAPTER ONE

Dealing with Darwin:
Locating Encounters with Evolution

T HOUGHTS TRAVEL. But as they journey around the world they do not
move effortlessly from place to place, from site to site, from setting to
setting. In different venues they mean, and are made to mean, different
things. This is because the circulation of ideas is not simply about transference;
it's about transformation. It's not just about dissemination; it's about appropriation. In 1982, Edward Said called attention to this phenomenon in a classic essay
entitled "Traveling Theory." His point was simple but profound: as theory travels
it transmutes. The movement of any theory into a new space, he observed, "necessarily involves processes of representation and institutionalization different
from those at the point of origin." And this "complicates any account of the trans-
plantation, transference, circulation, and commerce of theories and ideas."[1]

Since then this insight itself has traveled in many different directions, not
least into the realm of scientific inquiry. Thus James Secord has suggested that
we need to begin to "think about knowledge-making as a form of communicative
action." We have "to recognize that questions of 'what' is being said can be an-
swered only through a simultaneous understanding of 'how,' 'where,' 'when,'
and 'for whom.' "[2] With science, as with other spheres of knowledge, migration
is not simply about mobility; it's also about mobilization. For as scientific theory
is encountered in different locations, it is translated into different idioms, fashioned into different forms, and harnessed for different purposes.

Charles Darwin's theory of evolution has moved around the world. And different communities have had different dealings with it. My quarry in this book is to
track down something of how Darwin was encountered, what he was understood
to be saying, and what his theory was seen to signify in a range of different settings in the decades around 1900. More specifically, the aim is to inquire into how
communities sharing the same confessional heritage dealt with Darwin in different
places and, at the same time, to inspect the kinds of deal they struck with evolutionary theory in their own situations. My strategy is to focus on a suite of fairly
specific venues where evolution was the subject of discussion and debate. In part
what motivates this chase is my allergy to "isms" of various stripes—Darwinism,

evolutionism, creationism, and the like. There is, of course, a brisk trade in such bloodless abstractions. Claims to inevitable warfare between science and religion, or evolutionary theory and religious conviction, abound and dominate the public sphere. Less commonly, declarations of intrinsic harmony can also be heard. But these portrayals are only sustainable at the cost of essentializing either "Darwinism" or public reactions to it. Generalities of that stripe, common though they are, only serve to substitute the monochrome of caricature for the Technicolor of real life and to favor the clean, crisp lines of presumption over the messy particularities of history. As I read the historical record, by contrast, I find complexity and contradiction, contingency and complication that defy simple typecasting. For in every case, I believe, the encounter with Darwin needs to be understood in terms of domestic circumstances.

What also animates my mission is the conviction that there is much value in attending to what I call the "geographies of reading" and the dynamics of "speech spaces" in making sense of cultural engagements with Darwin's theory.[3] By "geographies of reading" I mean the different ways scientific proposals are read in different venues and how they are marshaled in particular places for particular projects. By "speech spaces" I refer to how specific venues condition what can and cannot be said about new knowledge claims, how things are said in those settings, and, just as important, how they are heard. Location and locution are intimately intertwined.

Text and Talk

Something of the value of elucidating the geographies of reading can be gleaned by calling attention to the cultures of book reviewing. Because reviews stage-manage texts and steer readers in particular directions, they actively participate in determining the significance that any work comes to hold for a reading community. Take the case of Alexander von Humboldt. As Nicolaas Rupke shows, French and German reviewers of Humboldt's *Essai Politique sur le Royaume de la Nouvelle-Espagne* tended to focus on how his new determinations of longitude, latitude, and altitude delivered increased cartographic accuracy. The British, by contrast, dwelt on its more commercial and strategic possibilities and how the work meshed with natural theology. Cultural preoccupations, in other words, channeled reviews in particular directions. Mapping these geographies of textual reception becomes a way of disclosing what Rupke calls "the constitutive significance of place in the production of the various meanings that become attached to even a single work."[4]

James Secord's analysis of the different ways the anonymous *Vestiges of the*

Natural History of Creation, a work of speculative natural history and evolution first published in 1844, could be read opens up further dimensions of the role of location in textual encounter. Embraced by some, vilified by others, *Vestiges* at once bemused, infuriated, consoled, and revolted readers in its bold portrayal of the drama of evolution. One thought it a "priceless treasure," another dismissed it as materialist "pigology." Some found it manly; others were sure they could detect a womanly hand behind its anonymity. For some it was daring; for others melancholic. Aristocratic readers found it poisonous while progressive Whigs thought it visionary. Moreover, reviewers outdid one another in the metaphors they devised to orchestrate the text for readers. The book's striking red binding prompted one to "attribute to it all the graces of an accomplished harlot." Yet another exclaimed, "Unhappy foundling! Tied to every man's knocker, and taken in by nobody; thou shouldst go to Ireland!" The book's anonymity, moreover, generated a distinct geography of authorial speculation. As readers wrestled over how to fix an ethereal author, it mattered whether the writer was from the gentry or the working class, was a believer or an infidel, was a gentleman or a cad. Names that circulated in Edinburgh rarely surfaced in Oxford, while "those that were common in London's fashionable West End were barely known in the Saint Giles rookeries only a few blocks away."[5]

Reconstructing the geography of reviewing cultures constitutes only one route into thinking about the spatiality of reading. Ideas and imaginings, movements and methods, traditions and reputations all display their own geographies. Examples could readily be elaborated. Diarmid Finnegan, for instance, has uncovered something of the different ways in which the glacial theory was interpreted in a range of civic and cultural spaces during the Victorian period by tracking its fortunes through diverse sites and communication circuits—correspondence networks, lecture halls, science journals, and field sites.[6] On a different scale, we are now aware that such grand intellectual movements as "The Scientific Revolution" or "The Enlightenment" are better thought of as imagined singularities— abstracted fabrications—that collect under unitary labels markedly different projects and practices in different localities.[7]

Recent ventures in the field of biography have also gestured toward what might be called "reputational geography." Biographers of Isaac Newton in the years between 1820 and 1860, for example, when his reputation as a genius was invented in the public imagination, typically used him in ways that chimed with their own social, moral, and scientific values. Newton, to put it another way, was freighted with cultural baggage of one kind or another.[8] During the mid-1600s, Isaac La Peyrère put forward the simple but shocking thought that Adam was not the father of the entire human race and thus that humanity was of multiple, or polygenetic, origins. Later, during the nineteenth century, he was variously staged as an archetypal infidel, a far-sighted free thinker, and a cruelly maligned but profoundly orthodox believer. Whether he was cast as heretic, hero, or harmonizer crucially depended on the cultural location of the commentator in question.[9] And it was much the same with Alexander von Humboldt. At various points in German political history from late-Prussian times through the Empire Period, the Weimar Republic, the Third Reich, and the divided Germany of post-1949, Humboldt's identity was created and re-created to suit the political sensibilities of the moment. And so, Humboldt the liberal democrat, Humboldt the Weimar *Kultur* chauvinist, Humboldt the Aryan supremacist, Humboldt the antislavery radical, and Humboldt the pioneer of globalization were variously fashioned. Humboldt, Rupke writes, "has become a man with several lives, products of appropriation on behalf of geographically separate and chronologically successive socio-political cultures."[10]

Thoughts routinely travel the world in textual form, and the fact that print circulates opens up further possibilities for thinking about the geography of reading. Owen Gingerich's research on the marginal notes successive owners inscribed on the pages of Copernicus's *De Revolutionibus* enables him to track the fortunes of the new heliocentric theory in different settings. These jottings provide clues to how this canonical text was encountered in diverse locations at various points in time and constitute a kind of temporally extended set of reading clusters by which successive meanings were fashioned in dialogue with previous readers. As Gingerich concluded from his Herculean quest to track down every surviving copy of the text, the annotations collectively comprise "a precious legacy of the way in which the book was perceived and read during the scientific Renaissance."[11] In a comparable vein, Innes Keighren has traced out a book geography of Ellen Churchill Semple's *Influences of Geographic Environment,* which first appeared in 1911. What is striking in this analysis is that readers' reactions to hearing Semple speak in public often shaped how they read her book. What's no less significant is that while Semple herself shunned the term "geographic determinant" and claimed to only make mention "with extreme caution of geographic control," she was routinely cast as an environmental determinist *par excellence.* Whatever she intended to communicate, readers persistently took her to be saying something else.[12]

All of these inquiries serve to highlight the instability of scientific meaning and to demonstrate that while texts may be immutable mobiles—to adopt Bruno Latour's phrasing—their meanings are entirely mutable.[13] What further complicates things is that textual circulation often involves *literal* as well as metaphorical translation across contact zones of one sort or another. In her analysis of the cultural politics of late nineteenth-century scientific translations into Arabic, Marwa Elshakry directs attention to "the specific problem of rendering appropriate and meaningful lexical equivalents in cross-lingual scientific discourse" and the dilemmas it presents "for understanding the geography of knowledge, the way in which science travels across borders and through time."[14] Translations into Arabic of the term *science* itself, for example, were profoundly implicated in the politics of language since they abutted on matters of political power, religious orthodoxy, social change, and literary tradition. When we realize that there were no immediate Arabic equivalents for terms like *race, species,* and *evolution,* the dilemmas multiply alarmingly for translating a work like Darwin's *Origin of Species by Means of Natural Selection, or the Preservation of Favoured Races in the Struggle for Life* and dramatically bring to the fore what James Clifford pertinently refers to as the politics of neologism.[15]

The coming together of texts and readers, then, is a moment of creativity in which meaning is made and remade. For the encounter with words on paper is not to be thought of as a passive "consumption" of ideas; it is rather a positioned rendezvous, a situated dialogue, a sited engagement between text and reader. Acts of reading always involve *located* hermeneutics because readers are always part of what Stanley Fish calls "interpretative communities" sharing some foundational assumptions and exegetical strategies.[16] And this means we need to develop much greater sensitivity to the politics and poetics of place in textual engagement. So if, as Jonathan Rose puts it, "the history of reading is the history of interpretation," then geographies of reading turn out to be geographies of interpretation.[17]

No less important than text in intellectual commerce is another fundamental element in human communication—talk. And like textual encounter, talk is transformative. In a stimulating set of reflections on the art of conversation, Theodore Zeldin observed: "When minds meet, they don't just exchange facts: they transform them, reshape them, draw different implications from them, engage in new trains of thought. Conversation doesn't just reshuffle the cards: it creates new cards." What's important here is that talk is not simply about transfer; it's about transmutation. "Humans have changed the world several times by the way they have had conversations," Zeldin insists. "When problems have appeared insoluble, when life has seemed to be meaningless, when governments have been powerless, people have sometimes found a way out by changing the subject of their conversation, or the way they talked, or the persons they talked to.

In the past that gave us the Renaissance, the Enlightenment, modernity and post-modernity."[18]

The same transformation happened with science. It was the advent of a different *kind* of speech that stimulated the growth of European natural philosophy during the seventeenth and eighteenth centuries. New *ways* of conversing were ushered in. At the Royal Society of London, for instance, fellows were expected to talk plain talk, without frills or flourish. If people found the society's doings boring, well, that was precisely what they were meant to be. Francis Bacon thought everything should be done to eradicate from natural philosophy all "ornaments of speech, similitudes, treasury of eloquence and such like emptiness."[19] Brevity and concision were the virtues he prized. Changing the style and substance of the conversation had dramatic consequences. New forms of intellectual inquiry, new modes of thinking about the world, and revolutionary ways of conceiving the political order emerged. And the conversation spilled over into new spaces. In drawing rooms and workshops, in learned societies and dining clubs, in public houses and church halls, the new scientific conversation could be heard. In early eighteenth-century England, for example, the parlor became a new site of conversation about natural philosophy. When the first orrery made its appearance in 1716, it became a centerpiece of polite conversation.[20] Joseph Addison visited a household where he discovered women making jam and reading aloud from Fontenelle, the noted gourmand and philosophical author, on the usefulness of mathematical learning. In the kitchen he found them "dividing their speculations between jellies and stars, and making a sudden transition from the sun to an apricot, or from the Copernican system to the figure of a cheese-cake."[21]

All of this awakens us to the diverse arenas in which scientific discussion was conducted and to how different conventions governed different sites of talk.[22] Taken together these constitute distinct speech spaces where place and protocol combine to condition conversational exchange. The list of venues where the new scientific dialogue took place is extensive. The coffeehouse, for example, emerged as a site of scientific talk in the salon culture of Enlightenment Europe, and later, as the nineteenth century wore on, dining clubs, stately mansions, polite soirees, and laboratory tea rooms all played host to the new forms of colloquy.[23] Each venue had its own informal rules of conversational decorum and reflected social attitudes embodied in etiquette manuals and changing conventions about shoptalk.[24]

The intimately reciprocal connections between speech and space have far-reaching consequences. Speech spaces shape what can and cannot be said in particular venues, how things are said, and how they are heard. In different arenas there are protocols for speech management; there are subjects that are trendy and subjects that are taboo. In public spaces and in camera, in formal gatherings and in private drawing rooms, in conferences and consultations, in courtrooms and churches, in clinics and clubs—in all these venues—different things are speakable . . . and unspeakable . . . about scientific claims. Individuals moving between these spaces adjust their speech—code switching, as it is called—to suit the setting. In so doing, Peter Burke remarks, they are "performing different 'acts of identity' according to the situation in which they find themselves."[25] This means that the control of speech space is intimately connected with the maintenance of identity. Spaces of speech, of course, are also spaces of silence. There are always voices that are absent or not allowed to speak or denied access. In colonial societies, James Scott powerfully reminds us, the oppressed can rarely let their voices be heard.[26] No doubt for different reasons, but with not entirely dissimilar effects, those marginalized in scientific debates find their voices unwelcome in science's privileged sites.

It is not difficult to see how crucial speech spaces are, impinging as they do on matters of rhetoric, audience, tone, venue, reputation, and the like in encounters over religion and science. Take, for example, what might be called "family space." A sensitivity about what could be said at home was something that leading naturalists reflected on from time to time with their close associates. The botanist Joseph Dalton Hooker spelled out the dilemma in an 1865 letter to Charles Darwin in which, as John Hedley Brooke puts it, "the social pressures for conformity were perfectly explicit."[27] "It is all very well for Wallace to wonder at scientific men being afraid of saying what they think," Hooker wrote. "Had he as many kind and good relations as I have, who would be grieved and pained to hear me say what I think, and had he children who would be placed in predicaments most detrimental to children's minds . . . he would not wonder so much."[28]

Such concerns easily spilled over into anxieties about what a person was comfortable saying in more public arenas. Darwin found it "a fearfully difficult moral problem about the speaking out on religion."[29] This reticence, of course, makes it extraordinarily difficult to ascertain just precisely what he really *did* think about certain subjects. For, Matthew Day reminds us, the "penchant for deliberate and sometimes dissembling cultural self-fashioning could be particularly conspicuous when religion was the subject of conversation."[30]

In institutional settings, how scientific theories were talked about required special care. And the consequences of intentionally—or unintentionally—violating customary expectations could have far-reaching results, as Alexander Winchell discovered to his bitter cost. As he mused in an explanatory letter to the readers of the *Nashville American* on 15 June 1878: "I have always taken pains, in my lectures at Nashville, to avoid the utterance of opinions which I supposed were disapproved of by the officers of the University."[31] But he obviously didn't succeed in his tongue tactics. He lost his chair at Vanderbilt University over issues rotating around human evolution and race history. In Belfast, as we shall see, in the aftermath of the assault that the religious establishment had received from John Tyndall's taunting speech at the British Association meeting of 1874, speakers in pulpits, on platforms, and at presbytery meetings talked about little else. Many complained that what had catapulted him into public controversy was the way he had infringed the association's unwritten rules of oral propriety.

Indeed high-profile clashes, like the infamous altercation between Thomas Henry Huxley and Bishop Samuel Wilberforce at the 1860 meeting of the British Association for the Advancement of Science cannot be understood, in my view, without attending to whether or not rhetorical decorum was breached during the row. Matters of etiquette and good taste were certainly uppermost in the minds of some who reflected on the occasion. Frederic William Farrar, later Dean of Canterbury, recalled that what the bishop said was neither vulgar nor insolent but flippant, particularly when he seemed to degrade the fair sex by pondering whether anyone—whatever they thought about their *grandfather*—would be willing to trace their descent from an ape through their *grandmother*. In Farrar's opinion, everyone recognized that the bishop "had forgotten to behave like a gentleman" and that Huxley "had got a victory in the respect of *manners* and *good breeding*."[32] Yet, while later writers thought Huxley urbane, at the time both the *Athenaeum* and *Jackson's Oxford Journal* found him discourteous. The boundaries of civility shifted over the decades. So . . . did the bulldog bite the bishop, or did the bishop badmouth the bulldog? It all depends on the character of the speech space that late June afternoon.

Darwinism in Rhetorical Space

If these reflections on text and talk are in the right neighborhood, there are, I think, significant implications for understanding Darwinian dealings in different places. Just what Darwin was thought to be saying in different venues is fundamental to figuring out the terms different communities negotiated with evolution. Places, of course, are always in process and are not hermetically sealed from one another. They are also shaped by the actions of those inhabiting them. Places are not neutral containers, mere stages on which the dramas of life are played out. Rather, places are constituted by the activities of the actants occupying them; at the same time those doings are constrained by the systems of human interaction pertaining in specific locations.[33] No doubt there are interesting things to be said about these matters at the level of the nation state and about regional patterns of response to Darwin's proposals; indeed, there has been something of an industry in this branch of reception studies over the years. Two major publications in this vein warrant particular attention. Thomas Glick's *The Comparative Reception of Darwinism* dwells on how the *Origin of Species* was received in different countries—France, England, Japan, Mexico, the Netherlands, and so on—by national scientific institutions and intellectual traditions and its influence on state scientific research programs. The results are certainly rich, yet, quoting David Hull, Glick concedes in his preface that "no correlation seems to exist between the reception of Darwin's theory around the world and the larger characteristics of these societies."[34] In a comparable two-volume venture, *The Reception of Charles Darwin in Europe*—part of a series charting the European response to British and Irish authors—Glick and Eve-Marie Engels draw together a wide range of scholars scrutinizing the fortunes of Darwinism in different European states at different points in time. While still operating at "the broad level of description that taking an entire nation or culture as the unit of analysis requires," this project had the great merit of expanding the map of Darwinian encounters into less well-known terrain such as Finland, Estonia, Poland, and Lithuania. Right from the start the editors make it clear that "different people understood different things when they said 'Darwin' or 'Darwinism'" and rightly insisted that in many cases Darwinism "is ultimately a code word that stands for more than the particular set of ideas advanced by Darwin." Indeed. For many in the late nineteenth century, the label *Darwinism* was simply a signifier for a rather loose amalgam of evolutionary theory that incorporated the thinking of the likes of Huxley, Spencer, and Tyndall as well. This acknowledgment, alongside the realization that "detailed analysis of the individual country-specific situation reveals that simple and broad correlations are not possible," prompted them to acknowledge that "quite different strategies are required" to deal with Darwin's fate among religious communities.[35]

From an even higher altitude vantage point, one of the world's most-cited interdisciplinary science journals, *Nature,* ran a series of sketches in 2009 on the fate of Darwin in major world regions. Under the rubric "Global Darwin," it hosted essays with such eye-catching titles as "Eastern Enchantment," "Revolutionary Road," and "Multicultural Mergers." One writer dealt with what might loosely be called the Middle and Far East and sought to show how Darwin's ideas were accommodated to Confucian, Islamic, and Hindu cosmologies. In so doing these cultures could claim that "Christianity alone was in conflict with science."[36] Another looked at the situation in China, arguing that there "under the threat of

Western imperialism, interpretations of Darwin's ideas paved the way for Marx, Lenin and Mao."[37] Yet another called attention to Latin America, insisting that in this part of the world people "first saw evolution as a reason to 'whiten' their societies, then as a reason to take pride in their mixed lineage."[38]

Valuable though such broad-brush canvases undoubtedly are, my concerns here are with more specific venues of Darwinian debate and the constructions that were put on evolutionary theory in particular settings. Thumbnail sketches of a few sites from around the world should suffice to illustrate something of the way in which the meaning of Darwin's theory was inflected in different locations by local circumstances.

Consider first conditions at the Charleston Museum of Natural History in South Carolina in the years immediately following the appearance of Darwin's *Origin*. Among the noteworthy naturalists who congregated around the museum was one of the leading students of the *Hydrozoa* in North America—the mathematician and marine invertebrate naturalist John McCrady (1831–1881).[39] Mc-Crady remained a lifelong opponent of Darwin's theory. On the surface this is surprising because he devoted himself to the construction of what he called a "law of development by specialization" which he believed encompassed all life. In the early 1860s he even noted that Darwin had "furnish[ed] a most beautiful explanation of the *Modus operandi* which probably characterizes the law of development in the production of specific forms and varieties."[40] But McCrady always conceived of development in ways akin to embryological growth that retained the identity—not the transmutation—of the individual, and so he could never go very far with Darwin. Why? To answer that question takes us to the core of Mc-Crady's politics. He was a tireless apologist for southern culture, issuing periodic commentary on the deterioration of civilization in the northern states, and extolling the excellence of its southern counterpart. The only hope for the South was secession, for, as he put it on the eve of the American Civil War, "a slave State never can be a centre of that form of civilization which now flourishes in Europe and at the North."[41] When the war came he enlisted in the Confederate States Army and rose to the rank of major in the corps of engineers.

Always concerned to keep nature and culture in tandem, McCrady readily took to mobilizing his scientific law of development by specialization in the interests of an independent South and calling on geological metaphor to naturalize geopolitical disruption. "The great Confederate Republic, founded by our forefathers, is about to break up into two or more confederacies," he remarked in 1861 in an article tellingly entitled "The Study of Nature and the Arts of Civilized Life." The "separation of this Union will be a convulsion," he went on, "but, like those vast convulsions of geological times, it will be a convulsion of development—a pang and throe of the birth-time of great nations which are yet to be—a grand and majestic step in advance." Here political and natural history were at one, for, as McCrady insisted, "if this be the course of our development, then is it in perfect harmony with all other great developments in nature, proceeding as they do by a progress in specialization."[42] But Darwin's science was an entirely different matter. His ideas about human origins and species transmutation were profoundly troubling. For McCrady was dedicated to the idea of racial superiority and closely followed his teacher Louis Agassiz in insisting that the different races constituted different human species.[43] Each race had a separate point of origin, and any blurring of its transcendental individuality was both biologically and socially repugnant. McCrady thus repeatedly insisted that it was simply impossible to conceive that the white and black races could have descended from the same origin. To him, Darwin's theory of species transmutation was nothing less than a subversive threat to southern racial and religious culture. *That* was the meaning McCrady discerned in the *Origin of the Species.*

In McCrady's case, his reading of Darwin was shaped by the cultural politics of his interpretive community—the circle of natural historians that rotated around the Charleston Museum of Natural History, where scientific inquiry was routinely domesticated to the needs of the Old South.[44] There, figures like Edmund Ravenel, John Holbrook, Lewis Gibbes, and Francis Holmes cultivated a distinctive *style* of southern science. Conceptually, it was characterized by what might be called an acquisitive Baconianism in which collections took precedence over conjecture, specimens over speculation. Not surprisingly, the prevailing intellectual context within which natural history was prosecuted was derived from natural theology. Again and again, Charleston naturalists located their mollusks and their hydrozoans in the framework of divine providence. Yet even more diagnostic of Charleston science was the extent to which it was hammered out on the template of political ideology. And nowhere was this more clearly manifest than in their efforts to seek in natural history scientific justification for their ideas about social hierarchy, racial superiority, and the supremacy of southern civilization.[45] Ravenel, for example, declared that abolitionists could not obliterate the laws of nature. Holbrook and Gibbes actively connived with Samuel George Morton, the Philadelphia medical practitioner, to marginalize monogenist opposition to their thoroughgoing scientific racism. They all threw their weight behind Morton, Josiah Nott, and the scientific luminary from whom they drew inspiration, Louis Agassiz, in order to put racial polygenism on a firm scientific footing, to confirm that the black and white races were different biological species, and thereby to legitimize antiabolitionist sentiment. *This* was the textual space into which Darwin's work was cast. The meanings attributed to Darwin's theory among the Charleston interpretive circle were thus shaped by what was taken to be the theory's implications for racial politics, postbellum anxieties about the fragmentation of southern culture, sectional rancor, and attitudes toward the liberalizing politics of reconstruction.[46]

Things were very different half a world away in New Zealand, and so was the rhetoric surrounding Darwin's theory. Whereas the Charleston naturalists saw Darwinian evolution as subversive of racial hierarchy, in New Zealand it was welcomed as endorsing the runaway triumphs of white colonial settlement. A set of public lectures presented at the Colonial Museum introduced the residents of Wellington to Darwin's theory in 1868–69.[47] The speaker was the New Zealand politician William Travers (1819–1903), an Irishman from Limerick, botanist, lawyer, and correspondent of Darwin. In the *Origin of Species* he perceived a theory with immediate implications for race history. Just as the European rat, honeybee, goat, and other invader species had displaced their New Zealand counterparts, so the "vigorous races of Europe" were wiping out the Maori. It was an iron law of nature. In the struggle for existence, Travers insisted, whenever a "white race comes into contact with an indigenous dark race on ground suitable to the former, the latter must disappear in a few generations." Nor was this state of affairs to be lamented; to the contrary, it was to be embraced. Whatever the temporary moral disquiet attending the prospect of a culture's annihilation, he was sure that the historic successes of European culture meant that "even the most sensitive philanthropist may learn to look with resignation, if not with complacency, on the extinction of a people which, in the past had accomplished so imperfectly every object of man's being."[48] In reading natural selection through the lens of race relations, Travers was simply bringing to Darwin's text the long-standing colonial conviction that the Maori were fated for extinction by Nature. His encounter with Darwin's theory and the meanings he found in it were thus molded by the contingencies of settler-Maori politics and the desire to enlist enlightened science in the service of domestic colonial policy.

Travers was not alone. Other members of the Wellington scientific fraternity, no less schooled in the rhetoric of naturalized imperialism, were happy to confirm this construal of Darwinism. The Wellington medical practitioner and later member of parliament Alfred K. Newman (1849–1924), for example, in a statistical analysis of the "causes leading to the extinction of the Maori," called on the breeding researches of Darwin, Wallace, and Galton to underwrite his declaration that the "feeble" Maori were "dying out in a quick, easy way, and being supplanted by a superior race." It wasn't, he noted, cause for "much regret." To him inbreeding was the chief cause, but he was also sure that indolence was a contributory factor. It was the same in the United States, he insisted, where "negro slaves kept at work increased in numbers, whilst freed negroes steadily decreased."[49] Again, Walter Buller FRS (1838–1906), magistrate, ornithologist, and lawyer, used the occasion of his 1884 presidential address to the Wellington Philosophical Society to declare that aboriginal peoples must recede in the face of civilization. As "the Norwegian had destroyed the native rat, and as the indigenous birds and shrubs were being supplanted by the introduced ones," he mused, "so surely would the Maori disappear before the pakeha [white settler]." It was just an inevitability that the "aboriginal race must in time give place to a more highly organized, or, at any rate, a more civilized one." It was simply "one of the inscrutable laws of Nature."[50] Pronouncements of this stripe have led John Stenhouse, noting that the "scientific establishment . . . favored Darwin from beginning to end," to observe that "New Zealanders embraced Darwinism for racist purposes."[51] In this context, the idea of struggle as an irresistible primal force became the hermeneutic key to delivering a Darwinian apologia for *pakeha* politics. After all, Darwin himself was sure that civilized peoples were "everywhere supplanting barbarous nations."[52]

Elsewhere in New Zealand too, the supposed benefits of Darwinian struggle more generally were applauded. The editor of Auckland's *Southern Monthly Magazine*, for instance, declared in 1863 shortly after the appearance of *The Origin of Species*: "Whatever may be thought of Mr Darwin's views concerning natural selection and the origin of species, no one will be disposed to deny the existence of that struggle for life which he describes, or that a weak and ill-furnished race will necessarily have to give way before one which is strong and high endowed."[53] And the Scottish-born Duncan MacGregor (1843–1906), a radical evolutionist who occupied the inaugural chair of mental and moral philosophy at the University of Otago and later became inspector of asylums and hospitals for the colony, told the readers of his article "The Problem of Poverty in New Zealand" (1876) that natural selection had delivered to New Zealand's hardy colonists the possibility of wiping out those social misfits—the thriftless and the stupid—who should be prevented from passing on their "unspeakable curse" to offspring who could only be described as "the common sewer of society." The "hopelessly lazy, the diseased, and the vicious," who would once have been happily weeded out by the Darwinian law of natural selection, fell under MacGregor's frosty eye, and he bitterly grumbled that they were "eating like a cancer into the vitals of society." To him it was tragic that these "waste products" were no longer left alone by society to hopelessly struggle and die unaided. His policy was more in keeping with a coldly rational Darwinian logic: drunkards, criminals, and paupers should simply be incarcerated for the duration of their natural lives.[54]

The principle that made Darwin's theory attractive to these audiences—namely, struggle—was precisely what most perturbed the circle that gathered at the Saint Petersburg Society of Naturalists in late nineteenth-century Russia. Of central importance here were the interventions of Karl Kessler (1815–1881), first president of the society and professor of zoology at Saint Petersburg University from 1861. In 1879 he declared on Darwin's theory in an essay tellingly entitled "On the Law of Mutual Aid." Drawing on his research on the ichthyology of the Aralo-Caspian region and supplementing it with supporting evidence from the more general natural history of the Crimea, he condemned "the cruel, so-called law of the struggle for existence." Kessler certainly allowed that overpopulation could generate intraspecific competition for resources, but he was sure that Darwin had given way too much weight to it. The sciences of zoology and sociology, he believed, had ignored "the law of mutual aid, which . . . is if anything more important than the law of the struggle for existence."[55] He reported that he himself had witnessed the survival value of reciprocated care and cooperation among bees, beetles, spiders, reptiles, and a host of other creatures. In the human species, mutual aid undergirded society's material and moral progress.

Kessler's reading of Darwin did not remain an isolated textual event. It inaugurated a reading history that steered Saint Petersburg engagements with evolutionary theory. A number of Kessler's associates—such as Alexander Brandt, Mikhail Filippov, Vladimir Bekhterev, and Modest Bogdanov—were critical in the establishment of what Daniel Todes describes as the "Russian Mutual Aid Tradition."[56] Perhaps most visibly of all, Kessler's cooperative gloss was vigorously promulgated in the writings of the anarchist prince Peter Kropotkin (1842–1921), who achieved prominence through radical publications and political activism.[57] A member of the Saint Petersburg fraternity, he read the written version of Kessler's "mutual aid" speech and later published the book on which his scientific reputation largely rests, *Mutual Aid: A Factor in Evolution*. In grand cosmic style, he traced the principle of correlative sociability from its application in the animal world, through primitive human society and medieval urban life, up to his own day. Like the evolutionism of his Saint Petersburg associates, this was Darwinism with its Malthusian teeth extracted. That ideology had been confirmed through his own firsthand experience of the Siberian wilderness where, in the Vitim region, he engaged in zoo-geographical inquiries with Ivan Polyakoff. "We saw plenty of adaptations for struggling, very often in common, against the diverse circumstances of climate," he recalled. But at the same time, he went on, "we witnessed numbers of facts of mutual support, especially during the migrations of birds and ruminants; but even in the Amur and Usuri regions, where animal life swarms in abundance, facts of real competition and struggle between higher animals of the same species came very seldom under my notice, though I eagerly searched for them. The same impression appears in the work of most Russian zoologists, and it probably explains why Kessler's ideas were so welcomed by the Russian Darwinists, while like ideas are not in vogue amidst the followers of Darwin in Western Europe."[58] Kropotkin later put his finger on the nub of the issue in a letter to Marie Goldsmith: "Kessler, Severtsov, Menzbir, Brandt . . . and finally myself . . . stand against the Darwinist exaggeration of struggle within a species. *We see a great deal of mutual aid,* where Darwin and Wallace see *only struggle.*"[59]

These Saint Petersburg readings of Darwin, of course, were not conjured out of thin air but were fashioned by earlier textual encounters, notably with Thomas Malthus's theory of population. Both on the political left and right in Russia, Malthus's atomistic conception of society had already been castigated as a cold, soulless, mechanistic product of English political economy. What, Pyotr Nikitich Tkachev mused, could "the great Darwin" have had in common with Malthus, that "pastor-thief"?[60] On the radical wing, for example, Nikolai Chernyshevsy's antipathy culminated in his announcement in the late 1880s that the Malthusian struggle for existence that Darwin had lodged at the center of his biology made it impossible for many to see how the masses could be helped. Walter Bagehot's *Physics and Politics,* published in 1867 and translated into Russian in 1874, which applied Darwinian struggle to national life, particularly nauseated him and confirmed his thinking that cooperation rather than struggle must govern social affairs. On the political right, *The Origin of Species*—along with the works of Karl Vogt and Henry Buckle—had been banned by the Tsarists in the mid-1860s on account of what they took to be its revolutionary tendencies. This impulse found expression during the 1880s in the rejection of Darwinism by the naturalist and economist Nikolai Danilevsky largely on account of its use of the Hobbesian-style war of all against all which, in its Malthusian guise, he believed was "a reflection of English utilitarianism and competition."[61] Danilevsky's account was later taken seriously in ecclesiastical circles.[62] In an environment where the laws of nature and society ran in tandem, Darwin's theory of organic change was interpreted in the shadow of the Russian rendezvous with Malthus. Malthus might have been a credible intellectual source of inspiration for Darwin and his disciples in England, but his philosophy was profoundly troubling to a Russian society whose class structure and political traditions made it suspicious of competitive individualism.

Perhaps unexpectedly, neither the race question that, in one form or another, shaped the way Darwin was read among the cognoscenti in Wellington and Charleston nor the anxieties about struggle that motivated the Saint Petersburg naturalists, dominated the English-speaking conversation over Darwinism that surfaced in the *Cape Monthly Magazine* in South Africa's Western Cape during the 1870s.[63] The magazine had come into being in 1857 to advance the virtues of intellectual enlightenment, social progress, and the spread of civilization in the Cape. It occupied a strategic intellectual location at the intersection of the South African Public Library, the Art Gallery, and the University of the Cape of Good Hope.[64] Aspiring to involve itself in the global scientific conversation, its editors kept their eyes "firmly fixed on developments in the imperial centres of London and Edinburgh."[65]

In June 1874, the subject of Darwinism was stirred up by an address delivered at the South African Public Library in Cape Town. The speaker was Langham Dale, an Oxford-trained mathematician, superintendent-general of education, and a pioneer archaeologist.[66] While the tenor of his assessment was restrained, his misgivings about the Darwinian edifice were plain for all to see. Dale worried over natural selection's capacity to produce the first truly human pair and followed Alfred Russel Wallace's claim that the human mind was not subject to the operations of natural selection. But he fastened most comprehensively on the question of the evolution of language, insisting that there were "countless difficulties" facing any account that assumed that "interjectional utterances prompted by . . . sensations" could be the "germ of articulate speech."[67]

Dale's intervention provoked a number of responses in the weeks that followed, but whatever stance they adopted, their tone was conspicuously moderate. In July, an anonymous author contributed a piece on the origin of language, arguing that human speech could have developed by means of natural selection from the imitative articulations of prehuman hominids. By accumulative inflections and modulations, a rudimentary language could easily come into being with increasingly abstract ideas finding limited expression. Dale was simply "injudicious" in the stance that he had adopted.[68]

The most thorough interrogation came from Queenstown physician Sir William Bisset Berry, an Aberdeen-educated surgeon and later speaker of the House of Assembly.[69] In the 1874 August and September issues of the *Monthly,* he contested each of Dale's claims in painstaking detail. To Berry, evolution could comprehensively account for everything from species transformation and the emergence of mind to the development of sociality and the family, as well as moral sentiments and human language. On the issue of humanity's simian ancestry, he queried

Wallace's exemption of the human mind from the operations of natural selection and made much of John Fiske's data on human and gorilla brain size so as to net humans and primates together in an altogether compelling way. Berry confessed himself "astonished" by Dale's declarations on humanity's moral distinctiveness and, following John Locke, proclaimed the demise of "the doctrine of an innate moral sense." With that presumption disappearing into the abyss, he concluded, "the last stumbling-block of the anti-evolutionist" was whisked away.[70]

A few weeks later, in the October issue, Dale replied to his critics. His tone was firm but conciliatory. He briefly reiterated his thoughts about the immutability of the human species, the absence of transitional fossil forms, and his own sense that "a vital energy" pervaded the material world. Nevertheless, he chose to end his commentary with the words of John Tyndall: "I do not think that the evolution hypothesis is to be flouted away contemptuously . . . Let us hearken to those who wisely support it, and to those who wisely oppose it . . . The only thing out of place is dogmatism on either side."[71]

The Cape's former attorney general, William Porter, also offered moderately liberal commentary in 1876.[72] Retired to Ireland but in his capacity as first chancellor of the Cape University, Porter presented his inaugural address on the current intellectual scene. Born in Ulster into a New Light Presbyterian manse with strong liberal convictions, he worked tirelessly for equal justice for blacks and whites. In 1873 he returned to Belfast, where he lived with his brother John Scott Porter, a Unitarian minister, until his death in 1880. During these years he witnessed first hand the altercations over Darwinism that were stirred up by Tyndall's presidential address to the Belfast British Association meeting in 1874.[73]

While William Porter made it clear that many students of Darwinism did not perceive in it any pernicious impulses, he expressed considerable disquiet over the radically naturalistic Darwinism of Ernst Haeckel. Porter was certain he could detect "indications . . . in Professor Haeckel's book that his scientific creed has tended to pervert . . . his moral sentiments." What disturbed him most was Haeckel's heartless eugenic rhetoric. Killing off "all sickly, weak, and crippled children" so as "to promote the survival of the fittest" did not exactly warm Porter's heart. "Christendom," he proclaimed, "does not kill its sickly, weak, and crippled children. It builds hospitals for them." How different that creed was from Haeckel's Darwinian vision of a world characterized by "a pitiless and most embittered struggle of all against all"![74]

Porter's worries about these forms of radical Darwinism need to be placed in the context of his Unitarian liberalism. For his address was dominated by a pervading sense of the role of education in cultivating social cohesion, civic ethics, moral improvement, material prosperity, and intellectual capital. Anything that would sabotage these Enlightenment sentiments by substituting materialist values for moral virtue was politically—and religiously—intolerable. And that was precisely what Haeckel-style *Darwinismus* threatened to do—in the new University of Cape Town and in the Cape colony more generally. For someone powerfully animated by New Light liberalism on race relations, a loathing of oppression, and faith in progress, the dark face of Darwinism that Haeckel represented was nothing short of sinister.

Taken in the round, then, Darwinian conversation among supporters of the South African Library and the *Cape Monthly Magazine* during the mid 1870s was imbued with liberal sentiments. Support for the theory was judicious, criticism cautious. And the temperate tone that interlocutors adopted was entirely in keeping with the progressive, Enlightenment aspirations of the Cape's literati, whose eyes were firmly fixed on science's metropolitan horizon. The implications of Darwin's theory for local issues were not conspicuously paraded, but when his more radical champions pushed the theory in directions that threatened to undermine the civic, moral, and pedagogic progressivism at the heart of their liberalizing institutions and civilizing credo, urbane censure was certainly forthcoming.

Charleston, Wellington, Saint Petersburg, and Cape Town are only a few of the places where Darwin and Darwinism were talked about during the second half of the nineteenth century. Other sites have other stories. But these are certainly sufficient to demonstrate that in different venues the meaning of Darwinism was constituted in markedly different ways. The Charleston naturalists inspected Darwin's proposals through the lens of a racialized, Agassiz-type creationism that was more comfortably suited to the politics of postbellum segregation than evolutionary monogenism and for that reason remained unimpressed with his theory. By contrast, the colonial politics of New Zealand's settler society predisposed Wellington readers of Darwin to find in his writing a heartlessly robust selectionism that supported the cutthroat ethics of race and class struggle. In Saint Petersburg, Darwin's theory was construed in the light of a customary disquiet over Malthusian social theory and was translated into the language of reciprocal sociability, while the circle that rotated around the *Cape Monthly Magazine* read him against the backdrop of Enlightenment ideals, metropolitan ambitions, and liberal values. In different venues Darwin was enlisted in different campaigns; his name was recruited for different causes; his theory was made to stand for different political purposes.

Localizing scientific encounters with Darwin's theory in this way has major implications for charting the religious response to evolutionary theory. The instability of Darwinian meaning itself alerts us to the need to place encounters in their local settings. But it also calls attention to the role that cultural politics, in one shape or form, have played in how particular communities engaged with Darwin's novel proposals. What is just as clear, moreover, is that the rhetoric surrounding the Darwinian debates in different locations structured the encounter with evolution in profoundly important ways.

Religious Encounters with Evolution: Place, Politics, Rhetoric

If place, politics and rhetoric played such central roles in scientific engagements with Darwin, the self-same conditions surely etched themselves just as deeply into the fabric of religious responses to the theory.[75] Keeping a clear eye on local particularities opens up new dimensions of a subject too commonly buried beneath a veneer of presumption. Take, for example, Andrew Dickson White's staging of the geologist Alexander Winchell, who was dismissed from the Methodist-controlled University of Vanderbilt in 1878. Remarking on the campaign to stamp out evolutionary thinking in institutions under theological control, White told the readers of the *Popular Science Monthly*: "The treatment of Dr. Winchell at the Vanderbilt University in Tennessee showed the same spirit; one of the truest of men . . . he was driven forth for views which centered in the Darwinian theory."[76] Later, in his *History of the Warfare of Science with Theology in Christendom*, White elaborated on what he called this "disgraceful" history, explaining that Winchell's theory that humans had existed prior to the advent of the biblical Adam was a key component of the crusade against him.[77] Under the control of "one of the religious sects most powerful in that region," the institution had connived in charges characterized by "luminous inaccuracy" and dealt with Winchell in ways akin to Galileo's treatment at the hands of the Catholic church.[78] On closer inspection, however, we can see that place, politics and rhetoric were all implicated in the fracas.

To be sure, Winchell himself was convinced that his heresy consisted "in holding with the great body of scientific men, that a method of EVOLUTION has obtained in the history of the world," though he emphasized that he did not accept "that *man* is the product of evolution."[79] White was certainly accurate in identifying Winchell's pre-Adamism as a dominating concern. To grasp what was at stake, it is imperative to take into account earlier dealings with the new science of anthropology in the American South. During the 1830s and 1840s, figures like Samuel George Morton, George Gliddon, and Josiah Nott had vigorously promoted the idea that humanity was of plural origin and constituted a number of separate species. Their polygenetic thesis ran directly counter to the monogenism of the biblical narrative and was widely regarded as subverting the unity of the human race.[80] Their language, even for the time, was also viciously racist. Though he protested against the charge of polygenism, Winchell's style of speech sounded remarkably similar to these anthropologists. For him that point of genesis lay so far in the past that by the time the biblical Adam appeared on the scene, humanity had already diverged into a suite of permanently fixed racial types. To Winchell it was crystal clear that the black races, which he set out to establish as physically, psychically, and socially inferior to whites, were *not* descended from the biblical Adam but predated him. He insisted that because the question of pre-Adamite humanity was "a matter of scientific fact, we should unhesitatingly appeal to anthropology for a final answer," a stance that hardly endeared him to his fellow Methodists in the Vanderbilt community.[81] To them it seemed that he had placed the black races, and other non-Adamic humans, beyond the scheme of redemption.

After all, the Southern Methodists, though deeply committed to a paternalism that relegated members of the black race to a subordinate sphere, repeatedly expressed concerns for their spiritual wellbeing. In keeping with both emphases, they worked long and hard to establish an independent, all-black, Methodist Episcopal Church with which they sought to retain cordial relations. Winchell's pre-Adamism threatened to erode the basis of this settlement by bestializing the entire race. Declarations such as the claim, whether backed by science or not, that in every characteristic where "the Negro departs from the White man, he approximates the African apes" would do little to maintain fraternal union.[82] In the light of such announcements, southern commentators fastened on the irony in northern sympathizers' statements of support for Winchell and the virtues of scientific freedom. "It is certainly remarkable that these supereminent friends of the negro race . . . should be so very tender of a book that undertakes to prove that the negro does not belong to Adam's line," quipped one writer; another slyly remarked, "It may turn out that the vindication of the negro's right to a place in the Adamic household is committed to Southern people."[83]

As these interventions make clear, attending to the role of place, politics, and rhetoric in the controversy over Winchell exposes elements that the standard "enlightened-science-versus-religious-obscurantism" version of events entirely fails to capture. Courtesy of Winchell's pronouncements, the board of trustees at Vanderbilt University now occupied a rhetorical space in which evolution and polygenism were intimately intertwined. Racial polygeny became the lens through which that community's dealings with Darwin were refracted. As the editors of the *Nashville American* recognized, "'his development of evolution and polygenism' caused the abolition of his lectureship."[84] It was that intellectual marriage that was Winchell's undoing.

Location can work in other ways too by sculpting specific doctrinal commitments in particular settings. The religious response to evolution in Denmark is illustrative. Of critical importance here was the influential Danish theologian N. F. S. Grundtvig (1783–1872), whose followers—the Grundtvigians—constituted a significant element in the Evangelical-Lutheran Established Church. How they engaged with Darwin's theory was a product of the local theological culture in which the movement took shape. For one thing, while Darwin's theory was elsewhere often judged in the light of Paleyite purpose, for Danish Lutherans "natural theology constituted a blind alley for human rationality."[85] The writings of Immanuel Kant and Søren Kierkegaard were decidedly more prominent in their intellectual landscape. Crucial too in their attitude to evolution was the fact that while Grundtvig himself took the Genesis narrative literally, he was critical of those mainstream Lutherans who succumbed to what he considered to be a form of bibliolatry. Grundtvig, to put it another way, was literalist but not biblicist. To him the cardinal facets of the faith were the sacraments and the creeds; the Bible was relegated to a secondary role. Within the Grundtvigian circle these commitments spawned two very different reactions to evolutionary theory. The "orthodox" Grundtvigians, under the leadership of C. J. Brandt and adhering steadfastly to the literalist component, rejected evolution; by contrast, neo-Grundtvigians like Niels Lindberg and the even more liberal Valdemar Brücker, suspicious of a narrow biblicism and open to higher criticism, found the resources to embrace Darwin's theory. Hans Henrik Hjermitslev demonstrates how these local internal theological disputes and the stances that interlocutors adopted were bound up with "specific doctrines within local theological contexts," which, in turn, "had a crucial impact" on how the Darwinian challenge was handled. Because the theological rhetoric they deployed in their engagement with Darwinism "was a uniquely Danish phenomenon," Hjermitslev concludes that "the case of the Grundtvigians demonstrates that local contingencies made a crucial difference in mainline Protestant responses to Darwinism."[86] In Grundtvigian spaces, Darwin was talked about in markedly distinctive ways.

In late nineteenth-century Scotland, Darwin's name, as hero or villain, was marshaled for the wider cultural purpose of maintaining a sense of geopolitical identity. When Henry Drummond (1851–1897) faced a dozen presbyteries and synods in his own Free Church of Scotland charging him with heresy and urging that immediate action be taken against him in 1895, it looked like another plain case of the religious repression of evolution. Drummond, an evangelist and lecturer on natural science in the Free Church's Glasgow College, was well known for his extremely popular *Natural Law in the Spiritual World* (1883) and *The Ascent of Man* (1894), in which he brought an evolutionary outlook to bear on questions of morality, religion, and spirituality. His critics were incensed, and official protestations flooded in. What is conspicuous about these expressions of grievance, however, is their geographical origin: all came from Highland sources.[87]

Both Highlanders and Lowlanders believed that the line separating northern and southern culture in Scotland followed an equally firm division within the Free Church. One Gaelic-speaking Highlander, convener of the Free Church Highland Committee, remarked in 1862 that "the Christianity of the Highlands was and is characterised by peculiarities of its own."[88] A prominent Lowlander concurred, insisting that between "the Celtic, or rather Gaelic and the more Scottish sections of the Church is a difference of race" and adding that "such a natural racial divergence is, from its very nature, exposed to the poison of a spirit of suspicion and hostility." Within the Free Church, he went on, were "two races, two worlds" and, indeed, "many forces . . . tending to make almost two religions."[89] As time passed, the more decidedly and proudly conservative Highlanders felt increasing hostility toward what they regarded as a compromising southern liberalism.

Concessions on traditional forms of subscription to the Westminster standards, the inroads biblical criticism was increasingly making into the denomination's colleges, and other modernizing tendencies resulted in numerous complaints from Highlanders reaching the floor of the Free Church's general assembly. To them, the appearance of Drummond's *Ascent of Man* was only the latest in a whole string of disquieting compromises, and so they fastened on it as the occasion for urging, more strongly than ever, the case for secession. "Man-degrading" and "God-dishonouring," Drummond's volume presaged nothing less than "*disasters more ruinous*" than anything that had yet befallen Highland Calvinists.[90] Free Church Lowlanders didn't see it that way at all. Robert Rainy, the principle of the Free Church College in Edinburgh, sprang to Drummond's defense, as did other intellectual leaders of the denomination, as well as the most prominent Scottish newspapers, which cast Highland heresy hunters like Reverend Murdoch Macaskill of Dingwall as bigoted, bitter, and backward. The Gaelic-speaking Highlanders' demonization of Drummond and his philosophy was thus a major weapon in the arsenal of an increasingly desperate battle to retain their ecclesiastical and cultural identity in an era of rapid intellectual transformation and social change. And it can only have confirmed them in their judgments when their overtures were dismissed on the floor of the general assembly by 274 votes to 151.

Location then is not simply the backdrop against which, or even the stage on which, human affairs are transacted. Rather, it constitutes the meaning of historical events themselves. The furor over evolution that blew up in 1882 in an American college located in the Middle East serves as a final illustration. Once again, Andrew Dickson White announced the incident: "At multitudes of institutions under theological control—Catholic as well as Protestant—attempts were made to stamp out or to stifle evolutionary teaching. Especially was this true for a time in America, and the case of the American College at Beyrout, where nearly all the younger professors were dismissed for adhering to Darwin's views, is worthy of remembrance."[91] What had happened was that Edwin Lewis, a Harvard-trained geologist and chemist teaching at the Syrian Protestant College, had paid tribute to Charles Darwin in the college's annual graduating speech in 1882. Here he used Darwin as an exemplar of patient empirical inquiry and scientific excellence. Protests were hastily lodged, and the board of trustees in New York expressed its alarm. Younger colleagues and students expressed their support to no avail. Lewis was dismissed, and resignations from several colleagues quickly followed.

To get a clearer sense of the cartography of the whole issue, we must recall the deeper cultural and historical context.[92] The roots of the college itself were embedded in earlier American missionary interests in the Middle East. As part of its operations, scientific subjects were increasingly promoted as a means of demonstrating the superiority of Protestant rationality over what was considered to be a more superstitious Islam and, indeed, Eastern church. Accordingly, over the years, such "enlightening" disciplines as natural philosophy, geography, astronomy, and logic were integrated into the curriculum. Missionizing and modernizing clearly went together. By midcentury however, dissenting voices were beginning to be heard. Some worried that the spiritual thrust of the whole enterprise was being sacrificed on the altar of this-world preoccupations. To such critics, Christianity simply had nothing to do with steam engines, electric wires, railway lines, or telegraph cables. By contrast, however, scientific and technological subjects were in increasing demand from local students, not least Muslims, and as a consequence the college, officially chartered in 1866, rapidly expanded during the second half of the century. Part and parcel of this success lay in its diffusion of the latest scientific knowledge in Arabic translation, not least through a number of weekly and monthly periodicals. From the college's perspective, science was intended to be a key component of evangelism, and so the rhetoric of the compatibility of divine revelation in the Book of Scripture *and* the Book of Nature was on the lips of many teachers. For consumers, however, it promised rather more mundane advantages.

Following in the footsteps of his Harvard master Asa Gray, Lewis believed he was simply demonstrating how God worked through evolution in the natural world. Others didn't see it that way. To the trustees, Darwinism meant rank materialism and an ape ancestry. The speech, particularly when it was translated into Arabic and published in the college's monthly journal *Al-Muqtataf,* sent shock waves through the board, reinforcing the anxieties that had arisen after the magazine had run laudatory pieces on Darwin earlier in the year immediately after his death in April.[93] Lewis's observations fully confirmed senior missionaries' and board members' suspicions about the folly of the way the college had been developing for years. For them, it was madness to confuse enlightenment with conversion. At the same time, and all too ironically, the Lewis incident served to further the dissemination of evolutionary theory in the Arab-speaking world. And it afforded the Syrian scholar and Sufi Husayn al-Jisr the opportunity to argue that Darwin's theory could certainly be aligned with Islam and therefore, as Marwa Elshakry puts it, "to assert the superior rationalism of Islam over other . . . more dogmatic faiths, particularly Christianity."[94]

It turns out, then, that the significance of the Lewis spectacle cannot simply be reduced to another case of science versus religion. The interweaving of speech, space, and geopolitics reveals something of how the Darwin issue could become the means of hammering out all sorts of other issues. In this case, these included the links between social reconstruction and spiritual regeneration, the role of science in evangelistic projects, the opportunism of Islamic apologists, and the differing meanings that could be attached to Darwin in an Eastern Mediterranean setting.

Dealing with Darwin

These few stories reveal something of the different ways in which place, cultural politics, and rhetorical style matter in Darwinian deliberations among religious communities. For the Methodists at Vanderbilt during the late 1870s, it was just very difficult to think about Darwin in isolation from Alexander Winchell's deeply troubling connection of evolutionary theory to racial questions. Despite Winchell's protestations to the contrary, his writings were too redolent of the rhetoric associated with heterodox anthropological polygenism to secure the support he needed to retain his chair. Among the Grundtvigians in Denmark, location worked differently. Here it was the cultivation of a specific set of theological convictions about scripture, natural theology, and experience that shaped the local conversation with Darwinian biology. In late Victorian Scotland, Henry Drummond's promotion of a Darwinized theology afforded another occasion for Free Church

Highlanders—already exasperated by the inroads modernity was making into their denomination—to lodge a series of protests about the reckless direction in which the church was hurtling. In this clash the fortunes of Darwinian evolution were umbilically connected to matters of cultural identity. In early 1880s Beirut, the American missionary fraternity, already unsettled by the course the Syrian Protestant College was embarked on, fastened on Edwin Lewis's concessions to Darwin in order to challenge the new—and to them newfangled—wisdom on the role that science and technology should play in the curriculum of a missionary-based American college.

In all of these venues, delving into local culture and conditions exposes new dimensions of the evolution-religion interface. Episodic though such inquiries are, they go quite some way to alleviating the problematic resort to those abstract "isms" to which I referred at the start of this chapter. And yet there is, I think, a need to go beyond these one-off accounts to a rather more systematic interrogation of place, politics, and rhetoric in religious encounters with evolution. That is this book's ambition.

In order to begin the task of cultivating a geography of science-and-religion, my strategy here is to take one spatially distributed but consciously self-identifying confessional family—Scottish Calvinists—and to trace how in a range of different spaces that community confronted Darwin. Wherever they were located, they all exhibited an enthusiasm for their Calvinist heritage, taught from the same classic texts, and shared a seriously similar theological architecture. What united them was their subscription to the 1646 Westminster Confession of Faith, the use of the same shorter and longer catechisms, the practice of church government by elders (presbyters), and a lineage whose origins can be traced back to the Scottish Reformation. And yet, as I shall show, they responded in markedly different ways to what they took Darwin to be saying.

I begin with Edinburgh in the 1870s, focusing on how a late Enlightenment intellectual heritage combined with profound anxieties over a set of dramatic local challenges, particularly from the controversially brilliant William Robertson Smith, to shape the way Darwin was read. I then move across the Irish Sea and examine the influence that the infamous "Belfast Address" of John Tyndall, president of the British Association for the Advancement of Science, had on how Darwin was staged and spoken about among Presbyterian intellectuals in that city. Along the way I pause to compare the rhetorical space they constructed with how their counterparts in Londonderry and Dublin wrote and talked about Darwin. Toronto is the next port of call, in particular the Calvinist community congregating around Knox College. Here, in a setting where the kinds of public spectacle to which Edinburgh and Belfast were host was conspicuously absent, dealings with Darwin were critically connected with a renegotiation of Baconian induction. This shift resulted in a dialogue with evolution very different from what occurred in the more conservative scientific community and different indeed from the story of Darwin's fate in Montreal. In Columbia, South Carolina, southern Presbyterians engaged in a protracted campaign over James Woodrow's future as an advocate of evolution and a teacher in the denomination's seminary. Issues revolving around race relations, fears of northern infidelity, skepticism, and abolitionism, together with a longing for the glories of a bygone era, were all folded into a fracas that made headline news in the *New York Times*. Finally, we turn to Princeton, New Jersey, and the remarkably influential set of intellectuals there who came under the influence, in one way or another, of James McCosh and Charles Hodge. The shape of that dialogue over Darwin was set in part by the different rhetorical spaces that these figures occupied—one at the university, the other at the seminary—as well as by the distinctive version of evolution that major scientific figures at Princeton were championing.

I have entitled this study "Dealing With Darwin." The double meaning is entirely intentional. On the one hand, I am concerned to show how Calvinist communities in different cities *dealt* with the Darwin phenomenon—some rejecting it outright, others tolerating it, yet others embracing it. At the same time, I am interested in exploring the different *deals* these communities struck with Darwin in order to maintain fidelity to their own traditions, bearing in mind all the while that Darwin's name often stood for a more general, multifaceted evolutionism. On both counts, I will insist, place, politics, and rhetoric were decisive in how the encounter was conducted and how evolution was judged in these different venues.

CHAPTER TWO

Edinburgh, Evolution, and Cannibalistic Nostalgia

Robert Rainy (1826–1906) was the undisputed leader of the Free Church of Scotland and, for more than a quarter of a century, widely acknowledged as its elder statesman. In 1874 he was appointed principal of New College in Edinburgh—the Divinity Hall of the Free Church—where he had already served as a professor since 1862.[1] As he pondered what subject he should tackle for his inaugural address to the students that October, he hit upon one of the hottest topics exercising the Victorian religious mind—"Evolution and Theology." Given the prominent position he now occupied and the public role he was increasingly assuming within his denomination, his thoughts on the subject can be taken, to some degree at any rate, as emblematic of how the intellectual leadership of Calvinist Scotland dealt with the Darwin dilemma.

Rainy's final verdict was as crisp as it was clear. Biologically, evolution was theologically irrelevant. "Evolution is continually going on before our eyes" he declared, and even if "the evolution of all animal life in the world shall be shown to be due to the gradual action of permanent forces and properties of matter," it would have no bearing whatsoever on the "argument of the Theist." To be sure, Rainy did not rule out divine intervention in accounting for human origins, but by maintaining that the essence of the human species did not lie in the arrangements of its physical frame, he found it possible to liberate Christian anthropology from detailed questions over similarities in the human and anthropoid skeletons. Whatever "infidel tendencies" might be detected in some contemporary champions of evolution, Rainy insisted that he did "not regard the question, whether man's animal constitution could conceivably be developed from lower forms, as one of any great theological interest."[2] This certainly did not mean that there were absolutely no grounds for concern. The application of evolutionary mechanisms to "the development of man's mental, moral, and spiritual nature" and—most of all—the "assertion that Christianity can be accounted for, as a step, or a set of steps, in the natural development of the human race, under the operation of its permanent causes and laws" were entirely different matters.[3] But, if the inherently cautious and quintessentially political Rainy is anything to go by, neither the principle of species transmutation nor the processes of human genesis were truly bothersome to critical voices in Scottish Calvinist culture. What unsettled was the application of evolutionary naturalism to the structural development of Judeo-Christianity and its scriptural record as a cultural formation.

Because Robert Rainy worked hard at maintaining close links with the traditional heritage of the Free Church even while trying to keep abreast of contemporary intellectual life and because he was ever attentive to the sensibilities of his constituency, it is particularly significant that he felt able to endorse an evolutionary reading of human descent in his inaugural lecture that winter. That such views could be promoted at New College, the church's intellectual headquarters in Edinburgh, is evidence of a general lack of anxiety about the origin of species by evolutionary means among Scottish Calvinist leaders by the mid-1870s.

Early Encounters

This relaxed attitude had not always been the case. Early appraisals of Darwinian evolution in Calvinist Scotland were markedly critical. Within a year of the appearance of the *Origin of Species*, John Duns (1820–1909), a Free Church minister who in 1864 would move to teach natural science at the denomination's New College, snapped that until "some tapir" should be "caught in the act of becoming a horse" Darwin's theory would remain as empirically unconvincing and conceptually shallow as the anonymous *Vestiges of the Natural History of Creation*. To Duns, all the evidence of geology testified to the permanence of species whose existence was due to direct divine intervention.[4] A couple of years later, when he brought out his two-volume *Biblical Natural Science*, which purported to explain all the scientific references in scripture, he continued the attack, concluding that "species have a real and permanent existence in nature." Theories of development in general and the Darwinian version of evolution by natural selection in particular were "wholly opposed to the utterances of the Bible on these topics."[5]

What emboldened Duns's resolute biblicism was an intuitive sense of the rightness of the monogenetic account of human origins that the Mosaic narrative supported. Until "its insecurity is clearly and triumphantly shown," Duns insisted, the unity of the human race had to be retained at all costs. But that wasn't difficult. Following the well-worn path of Johann Friedrich Blumenbach, James Cowles Prichard, and R. G. Latham, he found a perfectly good explanation of human variety in the "power of *habitat*, climate and the like." Talk of multiple creations of the different races was entirely gratuitous. "Is there greater unlikeness between the head of the negro, the aboriginal Australian, the European, and the Hindoo," he rhetorically asked, "than there is between the head of the greyhound and that of the mastiff or the bulldog?" Obviously, no. All were varieties within a single species. As for human language, Wilhelm Humboldt, Max Müller, and Baron C. C. J. Bunsen all confirmed that linguistic variations had sprung from a single source. The sciences were on the side of scripture, as was morality. Both stood against the ugly polygenism of the Morton-Nott-Gliddon brigade and their acolytes in the American South, where the idea that different human races had separate origins had been used by some anthropologists and natural historians to justify slavery. It was only to be expected from that quarter, he slyly jibed, that in their portrait of universal race history they "would assign a foremost place to the families to which they themselves belong." But they might have achieved this self-designated distinction, Duns added, "without attempting to un-soul others farther removed from the early centres of civilization."[6] No more attractive were the dark evolutionary musings of Carl Vogt and the ethical relativism of the continental materialists. Both scripture and science united in their opposition to all these modernist subversions of human dignity and public morality.

Duns retained his early hostility to Darwinism for the rest of his life. By 1877, for example, he was still declaring that evolution was distinctly harmful to theology,[7] and later still he repudiated the efforts of those like Asa Gray who worked hard to devise a Darwinian teleology.[8] David Brewster was no more sympathetic. Principal of the University of Edinburgh since 1859, distinguished experimental physicist, and an elder in the Free Church, Brewster found way too much "wild speculation" in the *Origin of Species*. Certainly there were novel facts, but much of the book was composed of "little more than conjectures." It left a bad taste in Brewster's mouth, for he was sure that the whole "tendency" of the work was to "expel the Almighty from the universe, to degrade the god-like race to which he has committed the development and appreciation of his power, and to render revelation of his will an incredible superstition." Besides all this, much of it went against good science—against what was known about the permanence of species, reversion to type, and the fossil record. Frankly, Brewster concluded, Darwin had "not adduced a single fact" in support of his reckless speculation.[9]

No doubt a good deal of the early antagonism to Darwinism was stirred up by Thomas Henry Huxley's visit to Edinburgh's Queen Street Hall to present two lectures on Darwinism in January 1862. Before a packed audience, he attacked the biblical account of creation and declared that humans were descended from the same stock as apes. That threw the Free Church *Witness* into a spasm. Huxley, its readers learned, had promulgated the "vilest and beastliest" of doctrines.[10] For his part, Huxley was mightily pleased, reveling in the "large & liberal cursing" he had received from the Free Church coterie and adding, in a letter to Henry Dyster, that "Life has its joys, my son, if we earn them!"[11] Just a few months earlier, in October 1861, the controversial and colorful explorer Paul du Chaillu also washed up in Edinburgh to talk in the Music Hall about his African experiences.[12] Armed with a gruesome haul of disembodied heads and an arsenal of anecdotes, he titillated the audiences in his peripatetic tour of Britain with theatrical displays of stuffed gorillas and drew lurid parallels between apes and savages. Controversial and "disposed occasionally to paint things a little *couleur de rose*," according to Rev. Dr. Robson, du Chaillu was nonetheless commended by "a most excellent American missionary" as exhibiting exemplary "truthfulness, integrity, and Christian character."[13] Meanwhile, the Edinburgh novelist R. M. Ballantyne, author of *The Coral Island*, had drawn heavily on du Chaillu's *Explorations and Adventures in Equatorial Africa* for his 1861 novel *The Gorilla Hunters*.

The "monkey-to-man" theory—with its Darwinian odor—hung in the air.[14] And it remained firmly lodged in the memory of Rev. Thomas Smith, minister of Cowgate-Head Free Church, who, in a lecture to the United Presbyterian Congregation in Broughton Place in 1867, sneered at the idea of a gorilla transmuting into a human. Of all the "instances . . . to which the Darwinian theory of development is manifestly inapplicable," he urged, surely none was more clear-cut than the "question of the gorilla and man." "When we hear men labouring to prove the descent of our race from this disgusting baboon," Smith scoffed, "we always feel disposed to make use of an answer, which certainly is not quite philosophical, but which is particularly telling: 'Well! You may be a gorilla if you like, but I am not.'" To Smith, there simply was "not a single fact that goes in the direction of showing that a cabbage can be changed into a rose, or a horse into a cow, or a titmouse into an eagle, or a gorilla into a man." On the contrary, "innumerable facts" disclosed "insurmountable" barriers to transmutation.[15] No doubt such resistances drew comfort from the geophysical estimates of the age of the earth, based on the secular cooling of the globe, put forward in 1862 by William Thomson (later Lord Kelvin), who periodically worshipped at the Free Church in Largs.[16] By Thomson's calculations, there simply was not sufficient time for Darwin's mechanism of natural selection to have delivered the myriad life-forms currently exhibited throughout the natural world.

New Dealings

However unpalatable some Scottish Presbyterians found their first tasting of Darwinian evolution, rather less hostile assessments were already beginning to be heard. James David Forbes, for instance, reviewing Charles Lyell's *Antiquity of Man* in 1864, conceded that "man was certainly the contemporary of the large and well-known class of extinct mammalia hitherto regarded as belonging to an age of the world preceding his creation."[17] And in 1866, the Free Church editor of the *Daily Review*, David Guthrie, also spoke in favor of a progressivist rendering of evolution.[18] Soon other voices were added in support. John Tulloch, principal since 1859 of Saint Mary's College, Saint Andrews, made it clear in 1874 that he considered Darwin "a genius of observation," even if he thought him "as little a philosopher who ever lived." He told his readers that he had "no quarrel with the evolutionary hypothesis in itself"; to the contrary, it was "an inspiring conception."[19]

A couple of years later Robert Flint (1838–1910), who moved from the chair of moral philosophy and political economy in Saint Andrews to assume the professorship of divinity in Edinburgh University in 1876, was working hard to develop an evolutionary natural theology. Passionately concerned to demonstrate the rationality of religious belief, he urged that the "law of heredity," the "law of variability," the "law of over-production . . . which gives rise to a struggle for existence," the "law of sexual selection," and even the "law, or so-called law, of natural selection" could all be read as expressions of divine purpose and the manifestations of a deeper teleology. As he put it in his first series of Baird lectures in 1876, "the researches and speculations of the Darwinians have left unshaken the design argument. I might have gone further if time had permitted, and proved that they had greatly enriched the argument." Indeed, he reckoned that "Dr Paley would have held the design argument to have been in no degree weakened by the theory of evolution." Commenting on natural selection itself, he reflected that "some might even hold that design cannot be conceived of as realised in any other natural way." However individual instances of adaptation were to be explained, Flint happily noted that Huxley was entirely correct when he conceded that "the higher teleology" remained untouched by the mundane daily doings of evolution.[20] Of course, Flint was chary of the materialism with which "three generations of Darwins" had been associated, and he spelled out his objections in his second set of Baird lectures the following year, 1877, stressing both the impassable chasm between vegetable and animal life and the problems of reducing consciousness to material forces.[21] Nevertheless, Flint maintained that, while he had "challenged the theology of Mr Darwin," he had "no wish to dispute his science."[22]

It was much the same with the United Presbyterian clergyman Henry Calderwood (1830–1897). A supporter of the campaigns of Moody and Sankey in Edinburgh, an enthusiast for the Scottish philosophy of common sense, and a critic of William Hamilton's idealism, Calderwood took up the chair of moral philosophy in Edinburgh University in 1868.[23] Not surprisingly he had his reservations about Darwinism's capacity to account for human moral development, for in his mind it was simply mistaken to "argue from a theory based only on facts observed in the lower forms of animal life to the determination of the facts of moral life." The animal impulse toward self-gratification ran counter to the "law of *self-denial*," which Calderwood considered "an essential law of human life."[24] Besides, moving from fact to value—from observation to evaluation—was a classic case of the naturalistic fallacy, slipping from an "is" to an "ought."

Yet Calderwood found it possible to welcome evolution as a *bona fide* scientific theory. In 1881, for example, he observed that Darwin had "given fresh stimulus to thought, and . . . succeeded in gaining wide support to a theory of Evolution" and had vastly increased "knowledge of the relations and inter-dependence of various orders of animate existence." His account of evolutionary change, Calderwood affirmed, had "found a permanent place in biological science."[25] Thus when he spoke to the Royal Society of Edinburgh on the subject in January 1890, he conceded as "facts" species transformism and "Darwin's hypothesis of one or two primordial germs." With the principles of the "struggle for existence," "adaptation to environment," and "hereditary transmission" of variations, Calderwood had no quibble, though he did wonder how "nerve sensibility [could] provide for evolution of intelligence."[26] His *Evolution and Man's Place in Nature,* which came out in two editions during the 1890s, thus began and ended with affirmations of organic evolution, whatever he felt about its inapplicability to consciousness, morality, and intelligence. In the first chapter he noted, "Whatever limitations are to be assigned to the theory, we must at least grant that a law of Evolution has had continual application in the world's history." In the final pages he reiterated the point: "Evolution stands before us as an impressive reality in the history of Nature."[27] In the second, much reworked edition, he declared the evidence for "descent with modifications" to be "so abundant and varied, as to leave no longer any uncertainty around the conclusion that a steady advance in organic form and function has been achieved in our world's history."[28]

All this confirmed what Calderwood had announced in his self-consciously irenic Morse lectures in Union Theological Seminary, New York, in 1880, in which he castigated John William Draper's inclination to concentrate on extremists rather than moderates in his notorious *Conflict between Science and Religion.* Even if "the theory of the Development of Species by Natural Selection . . . were accepted in the form in which it is at present propounded," he declared, "not only would the rational basis for belief in the Divine existence and government not be affected by it, but the demand on a Sovereign Intelligence would be intensified." Here he assured his audience that Darwin's theory "is no more at variance with religious thought, than with ordinary notions of preceding times" and that "the fewer the primordial forms to which the multiplicity of existing species can be traced, the greater is the marvel which science presents, and the more convincing becomes the intellectual necessity by which we travel back to a Supernatural Intelligence as the source of all."[29] As he summarized the current state of play: "The scientific conception of the history of animal life is, that there has been a historical progression in the appearance of animals, in so far as lower orders took precedence of higher, while the higher have shown large power of adaptation to the circumstances in which they have been placed. In accordance with the whole principles regulating the relations of religion and science, religious men, scientific and non-scientific, will readily acquiesce in this modification of general belief, as largely favored by evidence which geology supplies, and supported by testimony drawn from the actually existing order of things."[30]

Given his typically Scottish Common Sense philosophical interests in mind and morality, it is understandable that Calderwood would also turn to the whole issue of the place of the human brain in the natural order. In 1879 he had produced *The Relations of Mind and Brain*, arguing that the brain was primarily an organ of sensory-motor activity, not thought per se. The relations between mental activity and physiology (unlike sensory-motor functions and physiology) were only indirect. But while he exempted human higher intellectual functions and moral sensibilities from natural selectionist explanations, he nonetheless emphasized just how closely the brain of "the human organism stands allied to lower orders of organism" and "how many homologies of structure there are." To him, human life and animal life were "constructed on a uniform plan." "Thus the brain," he concluded, "and the two sets of nerve lines, namely sensory and motor, are the same in nature and functions in all animals, from the frog to man inclusive, and they differ only in complexity of arrangement."[31] Whatever the ultimate differences, Calderwood clearly conceded the existence of significant continuities between human and animal neurophysiology.

Not surprisingly, when Calderwood participated in the 1884 third General Council of the Alliance of Reformed Churches in Belfast, he maintained the selfsame stance in a speech entitled "The Religious Bearings of the Doctrine of Evolution," in which he urged that Darwin's theory afforded an even grander conception of the designer than had hitherto been glimpsed.[32] His remarks were entirely in keeping with the series of sympathetic sketches of Darwin himself—whose modesty and powers of observation Calderwood greatly admired—that he penned for the *United Presbyterian Magazine* in 1888.[33]

Other prominent Scottish Calvinists made various conciliatory pronouncements on the evolution hypothesis during the 1880s. George Matheson (1842–1906), Church of Scotland theologian and clergyman first in Innellan on the Argyll coast and then in Edinburgh, and from 1890 a fellow of the Royal Society of Edinburgh, addressed what he called "the problem of evolution and revelation" in his *Can the Old Faith Live With the New?* (1885).[34] Extremist stances on Darwinism—either pro or con—perturbed him, and he sought a via media, "a meeting-place . . . between the old culture and the new." In large measure, Matheson's mission was to determine what implications, if any, evolution would have for Christian culture should its findings be verified. But the overall direction of his thinking was plain. He had "no hesitation in saying that the modern doctrine of evolution, and especially the modern doctrine as expanded by Mr Spencer, is more favourable to the existence of an analogy between the human and the Divine than any previous system of nature with which we are acquainted." Spencer's resort to the idea of some inscrutable power, some driving force—what he called "The Unknowable"—showed him to have transcendentalist sympathies very different from the materialist creed of certain ultra-Darwinian enthusiasts. Spencer, he declared, was fundamentally "a Darwinian *plus* a transcendentalist."

Matheson found all this extremely promising and convinced himself that he could discern the Judeo-Christian Creator lurking at the headwaters of the Spencerian system: "The God of Mr Herbert Spencer, far short as He comes of the idea of Christian theism, is identical with the God of Christian theism in this, that He is a Presence not outside of the world but in the world." With the omnipresence of such a Power, it was just as easy for the Christian "to admit that the man has grown out of the animal, as it is to hold that the man was made immediately from the dust of the earth." To Matheson there simply was no "incompatibility between the claims of evolution and the claims of creation." No matter where he looked—the origin of life, the condition of "Primitive Man," the operations of Providence, the evidence of design, the doctrine of incarnation, the principle of immortality, the spiritual life—when any of these were scrutinized in the light of evolution, Matheson could only come to one conclusion: "We have found that the old faith *can* live with the new." "If the doctrine of evolution should be proved to be the law of the universe," he went on, "Christianity will occupy towards it a closer relationship than that of mere adaptation; it will itself take its place as one of the main forces in the achievement of the process." Furthermore, "the only effect which the universal acceptance of evolution would produce upon the Christian claim . . . would be to rest upon a basis of science what has hitherto reposed only on a system of faith."[35]

All of this chimed with the analysis he had presented the previous year in an

article entitled "Modern Science and Religious Instinct," in which he insisted
that it was "to Mr Herbert Spencer that our age is chiefly indebted for the suggestion of a compromise between the claims of Science and the claims of Religion."
In this analysis Matheson set out to demonstrate that the three key elements that
constituted "the natural basis of religion—a sense of wonder, a sense of fear, and
a sense of dependence"—could all be understood from an evolutionary perspective. Intuitively, he acknowledged, those inclined toward an evolutionary account
were likely to consider such sentiments as the preserve of primitive humans
which would progressively dissipate in the higher stages of social development.
Matheson queried this conventional presumption, arguing that these three characteristics were much more likely to develop in the later stages of evolution since
they all presupposed a level of cognitive awareness about the natural world that
approached the scientific. A sense of reverence, the feeling of awe, and a consciousness of dependence all required a degree of rational reflection that went far
beyond the mere survival instincts of "primitive man." An evolutionary explanation, far from undercutting such sentiments or relegating them to some primeval
past, was thus the only way they could be fully understood. Matheson's conclusion was clear: "We deny that the doctrine of evolution is the antithesis to the doctrine of creation." To the contrary, "the process of evolution is really tantamount
to a process of continuous creation."[36]

As the nineteenth century wore on, Scottish Calvinist rapprochement with
Darwinian evolution continued to deepen and broaden. James Iverach (1839–
1922), Free Church Professor of Apologetics in Aberdeen from 1887 and later
principal, provided a perceptive evaluation of evolution in its broadest range in
The Ethics of Evolution Examined (1886) and *Christianity and Evolution* (1894). The
breadth of his scientific literacy was clearly on display as he engaged not only
with Darwin and Spencer but also the likes of August Weismann, George John
Romanes, Edward Bagnall Poulton, and E. Ray Lankester.[37] Critical to Iverach's
assessment was a settled conviction that it was "in the interests of theology to
welcome every conquest of science and every fresh proof of the universal reign
of law." This necessarily included the Darwinian theory of evolution, and Iverach
made it crystal clear not only that he accepted "the general course of the evolution
of the earth's history" but also that it was "no longer possible for us to think of
things and of life in the old fixed static way. The adaptations, the inter-relations,
the incessant movement of life revealed to us under the guidance of biologists are
simply marvellous." The mere fact of the differentiation of the human species into
a variety of race types was hard evidence of evolution in action. Even the Bible
taught "a doctrine of descent." "If all the races of men are modified descendants

of one primeval man," he explained, "and if descent with modification can account for all of them, where is the objection on Scriptural and theological grounds
to accepting a theory which simply extends to the whole world of organic life a
principle which theology has always contended for as true with respect to man?"
All in all there simply could "be no doubt that biologists have got hold of a most
fruitful hypothesis." To be sure, evolution was opposed to one "particular theory
of creation," but that theory—the "view . . . that each species or kind was directly
created by God . . . and has gone on reproducing itself after its kind"—simply
could "no longer be held."[38]

Iverach's enthusiasm for evolution, of course, was not without qualification.
He was troubled by the expansionist imperialism that afflicted too many of Darwinism's advocates when they insisted on applying its principles in realms far
beyond its appropriate sphere. To them, "Evolution must reign without a rival;
everything must bend to its sway." Such "imperious demands" were certainly
troublesome, Iverach conceded, but they "must not be allowed to . . . frighten
us away from the name, or to blind us to the truth which is contained in it." It
was the same with the issue of teleology. To him, those who insisted on evolution's fortuitous operations were just mistaken, for in profound ways evolution
was entirely compatible with teleology. Indeed he was of the opinion that Romanes's *Scientific Evidences of Organic Evolution* furnished "an argument for design that is much more magnificent than that based on special creation." What
leant support to this entire line of thinking was Iverach's sense that Darwinism
itself was littered with uncashed anthropomorphisms. The critical analogy that
Darwin had exploited between natural selection and the selective activities of
domestic animal breeders certainly provided significant insights, but there were,
nonetheless, real causes for philosophical concern. Wallace and Darwin, he contended, had really only succeeded in explaining "nature in terms of human nature." Darwin was forever going on about the capacity of natural selection to
"produce structures," and Poulton was far too inclined to anthropomorphize
natural selection. When challenged, such advocates typically responded that "the
language used is metaphorical." But Iverach would have none of that.

We have already had too much of the metaphorical in this department of science,
and the theory of natural selection has taken full advantage of what is merely metaphorical. It has grown to be a kind of *deus ex machina*, which seems to preside over
all changes of organisms, and . . . gives to the evolutionist all the advantages of a
presiding intelligence without its disadvantages. Natural selection is itself described
as a metaphor; but as soon as we begin to work with it its metaphorical character

disappears, and it becomes itself real, and is quite capable of anything. It has the character constantly ascribed to it both of a directing agency and of a presiding intelligence.[39]

What further reinforced this analysis was Iverach's keen appreciation of the degree to which the entire fabric of Darwinian language had been sociologically conditioned. Here he had learned much from the writings of his fellow Scotsman, the polymath Patrick Geddes, whom he considered "one of the most profound thinkers of our time." According to Geddes, modern evolutionary theory could be best characterized as the "substitution of Darwin for Paley." But this was not, as was often thought, "the displacement of an anthropomorphic view for a purely scientific one"; rather it was "merely the replacement of the anthropomorphism of the eighteenth century for the anthropomorphism of the nineteenth. For the place vacated by the logical and metaphysical explanation has simply been occupied by that suggested to Darwin and Wallace by Malthus in terms of the prevalent severity of industrial competition." Indeed, to Iverach, the inclination of too many evolutionists to naturalize struggle sprang from their eagerness to impute a capitalist mode of production to the natural order itself. This "reading of man's practices into the cosmos," he urged, was mistaken. For the most typical manifestation of the "struggle for existence" was not "cosmic" but human . . . all too human. "Unlimited freedom of competition, baker against baker, draper against draper, company against company, shipowner against shipowner, and one class against another," Iverach pondered, "thus we have the struggle for existence in its highest form." It was "the more virulent form of human competition" that had been installed as the driving force of Darwinian explanation. Not surprisingly, given the tenor of this rhetoric, Iverach was much more drawn to Peter Kropotkin's leftist emphasis on evolution by mutual aid.[40] Reciprocal sociability and communal care abounded in nature, he was sure, and he held out the hope that "as political economy is changing . . . and is learning to attach less importance to competition and more to co-operation, . . . those conceptions which biology has derived from political economy will also change . . . [T]he struggle for existence may neither be so keen nor so fierce as we have supposed it to be. We see in many cases that species, instead of striving for itself, may find its advantage in mutual co-operation."[41]

None of these assertions meant that Iverach was unenthusiastic for the insights of evolutionary biology. Indeed, he was willing to pursue its operations beyond the origin and transmutation of animal species into certain other domains. The idea, for example, of asserting "one origin for man's physical organism and another for his spiritual nature"—a strategy proposed in different ways by Wallace and St. George Mivart—was far too dualistic and thus deeply troublesome. Iverach thought that the "origin of man, body, soul, and spirit" had to be treated "as a unity." The idea of a soul being "superadded" to a human body was plain wrong-headed. Instead the entire human entity had to be understood as the product of continuous creation, namely, as the outcome of an evolutionary process ultimately guided by a supreme intelligence. He was also willing to extend evolution's reach into the realms of religious sensibility. Once "evolution as a method of the Divine working" was accepted, it was only reasonable to see "revelation itself" as critically implicated in the "process of evolution." Accordingly, "scripture itself is also an evolution, growing from small beginnings to greater and greater fullness and clearness." Such operations were part of a grand cosmic scheme: "The process of revelation was slow, evolutionary, progressive. Revelation was always related to the natural, proceeded on it, assumed, rectified it, and transformed it to a higher character and use. Even when, as Christians believe, revelation had been complete and redemption had been in essence realised in the work of Christ, then began a slow course of evolution, proceeding with many a backward curve, with many a sad reversion, yet on the whole upwards."[42]

While Iverach pondered these questions and, like Calderwood, sought to keep theology abreast of scientific developments, another Free Church writer was cultivating even more assiduously, and indeed expansively, a revisionist, Christianized Darwinism that has been dubbed "a thoroughgoing evolutionary cosmogony."[43] On completing his course at New College, Henry Drummond (1851–1897) was appointed in 1877 to a lectureship in natural science at the Free Church's College in Glasgow and promoted to a professorship there in 1883.[44] During these years, he accompanied the geologist Archibald Geikie (who had recommended him for the Glasgow post) to the American West and in 1883–1884 he traveled for six months in east central Africa. *Tropical Africa,* which came out in 1888, recorded his experiences and reported a number of scientific observations on geological features and tropical insects, notably the white ant. Throughout his short life, Drummond was also devoted to mission and happily allied himself with the evangelistic activities of D. L. Moody. While still a New College student, for example, he distributed leaflets for the Moody and Sankey Edinburgh mission, counseled penitents, and deputized for the transatlantic evangelists in various locations. He did the same thing during their missions to Britain in the 1880s and early 1890s and became one of Moody's closest friends.

In *Natural Law in the Spiritual World* (1883) and *The Ascent of Man* (1894), Drummond labored long and hard to elucidate continuities between the material and spiritual worlds. Like Calderwood and Iverach, Drummond was concerned to find in evolution the grounding for an ethics that gave pride of place to self-denial, altruism, and similar other-regarding sentiments. Throughout, a vibrant eschatological imagination animated Drummond's vision to unite the physical and metaphysical orders through the progressive evolution of philanthropy and self-sacrifice.[45] Precisely these predilections prompted one of his several biographers, James Young Simpson, to suggest that he "was not so much a biologist invading the world of religion as a poet invading and capturing the world of science."[46]

Natural Law in the Spiritual World, which began life as a series of lectures to working men in Glasgow's Possilpark, created a sensation when it appeared in 1883 and sold like hot cakes.[47] Moore estimates that by 1900 sales had risen to well above 120,000, and it had been translated into most European languages.[48] Its chief aspiration was to bring the principles of natural law into the spiritual realm and thereby to exhibit the profoundest continuity throughout the entire cosmos. The self-same laws—the laws of evolution—which governed the Polity of Nature, applied no less in the unseen State of the Spirit. As Drummond put it, "The position we have been led to adopt is not that the Spiritual Laws are analogous to the Natural Laws, but that *they are the same Laws.*" Religious sensibilities could no longer be considered the "Great Exception" to the embrace of scientific method, and he saw it as his task to demonstrate the merging of the natural and the spiritual in the outworking of universal law. If "Nature be a harmony," he contended, "man in all his relations—physical, mental, moral, and spiritual" must "be included within its circle."[49] His strategy was simple, if more than a touch whimsical. At base the book was an extended sermon or suite of meditations on how key themes in Christian experience could be expressed in the thought-forms of evolutionary biology.[50]

Thus the idea of spiritual life was couched in terms of biogenesis: "There are not two laws of Biogenesis, one for the natural, the other for the Spiritual; one law is for both." Similarly, the process of spiritual degeneration found its analogue in Darwin's elucidation of the law of reversion to type. Darwin's *Origin of Species,* Lankester's *Degeneration,* and Karl Semper's *The Natural Conditions of Existence,* he told his readers, were among the best works providing the "scientific basis" of the spiritual law of degeneration.[51] And so it went. Strong on conjecture, light on analysis, rich in allegory, *Natural Law in the Spiritual World* simply presumed the truth of evolution and worked at translating spiritual experience into biological language. So throughout, evolutionary motifs like heredity, environment, adaptation, unity of type, and habitat were called upon to elaborate laws of spiritual growth, eternal life, righteous living, spiritual death, mortification of the flesh, and the like.

A decade later, during the spring of 1893, Drummond presented a set of lectures, "The Ascent of Man," at the Lowell Institute in Boston. While the published book was extremely successful, it did not enjoy the widespread circulation of *Natural Law,* even though it has been described as his "greatest piece of work."[52] Drummond was now convinced that in at least one vitally significant way the nature of evolution had been "misconceived," and his lectures were his attempt to provide the antidote.[53] He promised to reveal to his audience evolution's critical "missing factor." Pretty well abreast of the contemporary literature at a time when Darwinism in its narrowest sense was in eclipse,[54] Drummond made it clear from the outset that the evidence for evolution—whatever might be the case about the omnipotence of natural selection—was "irresistible." He enthusiastically embraced it and indeed urged that its imperial reign should be extended to the widest possible compass. It "is essential," he declared, that evolution's "universal character be recognized, and no phenomenon in nature or in human nature be left out of the final reckoning." Even "moral facts" and "moral consciousness . . . must come within its scope."[55]

But, first, human physiological evolution required scrutiny. So he initially embarked on a review of the state of play in recapitulationist embryology, provided evidence of common descent from the anthropoid apes, and showed how the survival of vestigial organs was best explained on evolutionary principles. Along the way he paused to emphasize the critical importance of the struggle for life, natural selection, and the survival of the fittest in driving evolutionary history. So far, so good. But then, in his seventh lecture, Drummond moved to "a wholly new, and by far the most important, chapter in the Evolution of Man." Alongside the struggle for life, he introduced what he considered to be another foundational but largely ignored principle: "the struggle for the life of others." The transition from "self-ism to Other-ism," as he dubbed it, was "the supreme transition of history." The former was grounded in the demands of nutrition, the latter in the facts of reproduction. By providing a compelling account of the emergence of altruism, Drummond was certain he could open up the entire evolutionary process to spiritual values. Thus began his scrutiny of the role that "Other-regarding" sentiments played in natural history. Pride of place went to the self-sacrifice exercised by the mother in the care of her young and the significance of the prolongation of infancy in cultivating parental self-sacrifice and cooperation to ensure the survival and flourishing of offspring. It was clear that "without the Struggle for the Life of Others there can be no Struggle for Life, and therefore no Evolution." The critical principle that Darwin himself had elucidated required expanding if the full range of evolutionary explanation were to be understood. In the last analysis, evolution was "a doctrine of unimaginable grandeur" because it was "the Ascent of Love." The power of evolution to explain human physiology had been conclusively demonstrated. Now its *moral* force was coming clearly into view: "Viewed *simpliciter,* the Struggle for Life appears irreconcilable with ethical ends, a prodigious anomaly in a moral world; but viewed in continuous reaction with the Struggle for the Life of Others, it discloses itself as an instrument of perfection the most subtle and far-reaching that reason could devise."[56]

Having enthusiastically excavated instances of selflessness, altruism, and benevolence in the natural world—work which engaged with the chief proponent of mutual aid in evolutionary history, Peter Kropotkin[57]—Drummond's project reached its own Omega Point in the final pages of the *Ascent of Man.* Now his evolutionary cosmogony came into its fullest flower. He had not self-consciously embarked on any project to "reconcile Christianity with Evolution, or Evolution with Christianity," he explained. "And why? Because the two are one. What is Evolution? A method of creation. What is its object? To make more perfect beings. Through what does evolution work? Through love. Through what does Christianity work? Through love. Evolution and Christianity have the same Author, the same end, the same spirit."[58]

Not everyone shared Drummond's enthusiasm. Horatius Bonar, hymn writer and clergyman, wasn't impressed one little bit; "deadly poison" he snorted.[59] Across the Irish sea, Robert Watts—one of the "sleuth-hounds of heresy," to use Moore's choice phrasing,[60] to whom we will later return—mercilessly castigated Drummond's efforts as "unintelligible." They might have rhetorical charm, but they had no intellectual merit. "In a word," Watts concluded, "the casket is beautiful, embellished with all the grace and comeliness of a chaste poetic diction, but its contents mock our intelligence."[61] For his part, the Earl of Shaftesbury found Drummond's work "singularly pernicious."[62] Within his own denomination, a suite of overtures condemning *The Ascent of Man* was tabled by a conservative element from some Highland Presbyteries during the 1895 General Assembly, but a rebutting motion put forward by the New College principal, Robert Rainy, was carried by a majority of over 120.[63]

Still, according to Moore, outright rejection was never the typical response. Many preachers and evangelical workers found Drummond's writings to be a rich storehouse of sermonic analogy. To be sure, reviewers found flaws and quibbled about this or that, but by and large the thought that Drummond had creatively connected up God's world in a compelling saga of universal continuity gripped many readers. Indeed, his writings were widely admired and not infrequently interpreted as a courageous scientific apologia for faith, even if they fell short of demonstration. William Robertson Nicoll, long-time editor of the *British Weekly*, was sure that "the brilliant literary and theological renaissance of the Free Church" was largely due to Drummond and a few contemporaries, even if in private he thought Drummond—as he put it in an 1898 letter to Marcus Dods—something of a charlatan who was "always trying tasks far beyond him" and "as ill-read as a Bishop!"[64] The Oxford Anglican Aubrey Moore found *Natural Law* "remarkable" and "full of deep and original . . . thoughts," but hastened to say that in the last analysis it was "not really a great book, or one which will hold any permanent place in the history of apologetics" and "a brilliant example of a radically false method."[65] He seems to have been perturbed by two fundamentally countervailing tendencies he detected in Drummond's writings, namely, an unconscious inclination towards pantheism (evident in Drummond's resort to universal impersonal laws) and a more obvious predilection for Calvinism (revealed in his neglect of free will). Whether the accolades are justified or the censures warranted, however, both serve to confirm the main thrust of Drummond's project—the cultivation of an evolutionary theodicy. For, regardless of the accuracy of Drummond's scientific claims or the coherence of his theological proposals, there can be no doubting the depth to which the principle of evolution had penetrated into his mind and heart.

Just a year or two after Drummond's *Ascent of Man* appeared, Alexander Balmain Bruce (1831–1899), the Free Church's professor of apologetics in Glasgow, delivered his first series of Gifford Lectures. An outstanding scholar admired by German critics, Bruce was not enamored, as he himself put it, of "either the agnosticism of modern culture, or blind adherence to traditional dogmatism."[66] His 1897 Gifford series, *The Providential Order of the World*, simply presumed the potency of the evolutionary world picture and illustrated, in the early lectures, the extent to which it had penetrated into the heart of philosophical theology. It had, for example, transformed teleology, inducing "a great abatement in the confidence with which the teleological argument is regarded," though he was sure that ultimately evolution had actually enhanced the idea of purpose. "All may be mechanism," he insisted, "yet all may also be teleology." For himself, he was dubious about efforts—like Wallace's—to segregate the physical and the psychological, the material and the mental, and allocate only the former to evolutionary mechanisms. In contrast, he was convinced that "making man in his entire nature subject to evolutionary law . . . presents certain advantages for the cause of Theism." Both body and soul constituted an organic unity whose emergence he attributed to the operations of immanent divinity in the processes of evolutionary change: "man in all his characteristics, physical and psychical, is no exception to the universal law of growth, no breach in the continuity of the evolutionary process." It was simply mistaken to think that evolution debased humanity. "Evolution does not degrade man," he urged; rather, "man confers honour on evolution." Like Drummond—whose *Ascent of Man* he cited—Bruce could detect a gentler ethic in natural history than the brutalizing violence of Darwinism's harshest advocates and therein the grounds for the cultivation of personal, family, and social virtues. Furthermore, Bruce had come to the view that special creationism was a sort of heresy inasmuch as its presumption of sporadic interventionism tended to evacuate God, for the most part, from the universe. "The man who clings eagerly to the primitive impulse that set evolution going, to the origination of life, and to the inspiration of a living soul, as proofs that God exists, virtually declares that in all other parts of the history of the universe he finds no convincing evidence of God's being and power."[67] This idea was a clear echo of Aubrey Moore's dictum that "a theory of occasional intervention implies as its correlative a theory of ordinal absence." Indeed, to support his own viewpoint Bruce quoted Moore's observation: "Cataclysmal geology and special creationism are the scientific analogue of Deism."[68]

As the new century dawned and Darwinism narrowly construed came under continuing scientific pressure from a host of alternative evolutionary proposals—orthogenesis, the mutation theory of Hugo de Vries, neo-Lamarckism, and so on[69]—there were those like James Orr (1844–1913) at the new United Free Church College in Glasgow who registered to the full such reservations.[70] Orr, an Old Testament scholar and expert on the German theologian Albrecht Ritschl,[71] repeatedly argued that it was mistaken to equate Darwinism and evolution. His claim was certainly in keeping with the scientific spirit of the times. By the early 1890s he had revealed his attraction to DeVries's mutationism, which emphasized the role of saltations—the abrupt appearance of new organic forms—in evolutionary history.[72] Orr believed this theory overcame many of the problems in standard Darwinism, including gaps in the fossil record, reversion to type, and hybrid sterility. It had the added advantage of allowing him to read direct divine action into natural processes at critical points in organic development. It was particularly over the question of human origins that Orr had his gravest doubts about the all-sufficiency of natural evolution. In his 1903–1904 Stone Lectures at Princeton, he spent a good deal of time working to establish both physical and mental discontinuities between human and animal. At this precise juncture divine intervention was required. Evolution was not enough: "the balance of probability" was "in favour of man's exceptional origin." Nor was he drawn to the strategy that envisaged an evolved subhuman form receiving a human mind by an act of creation. "I confess it has always seemed to me an illogical and untenable position to postulate a special origin for man's mind, and deny it for his body ... Mind and body constitute together a unity in man ... You could not put a human mind into a simian brain." While for Bruce human psychosomatic unity encouraged a widening of evolution's role in human emergence, for Orr, it pointed toward a creationist resolution: "man, alike in his physical structure and in his spirit ... is not, as naturalistic theories assert, a mere product of evolution, but has, in a peculiar sense, his origin in a direct creative act."[73]

Nevertheless, whatever his reservations about Darwinian universalism, Orr had no doubts about the fact of evolutionary change per se. It was "not necessary to break with the general doctrine of descent to recognize the defects of the Darwinian presentation of it. Darwinism and evolution are not synonymous," he declared. What he referred to as a "considerable revolt" had "taken place in evolutionist circles from the idea of 'natural selection,' slowly operating, as the main factor in evolution." That shift had opened up fresh ways of thinking about organic change that found the driving force of transformation "*within* the organism, not in *external* causes."[74] All of this only conspired to confirm that evolution had taken place. "That species should have arisen by a method of derivation from some primeval germ (or germs) rather than by unrelated creations," he told his Princeton audience, "is not only not inconceivable, but may even commend itself as a higher and more worthy conception of the divine working than the older hypothesis." When God was conceived of as "immanent in the evolutionary process" he went on, "then evolution, so far from conflicting with theism, may become a new and heightened form of the theistic argument."[75]

Orr was even prepared to extend the range of evolution's explanatory scope into the sphere of theology itself. Thus he began *The Progress of Dogma* (1897) with the declaration: "I plant myself here, in truth, on the most modern of all doctrines—the doctrine of evolution, supposed by many to be fatal to the permanence of dogma. There has been evolution of doctrine in the past, and there will be evolution in the future. But evolution means that there has been something evolving; and *pro tanto*, if the evolution has not been utterly fatuous, there must be, as remarked about science, results of that process put on record." Indeed, he went on to apply the principle of natural selection to creedal formations in the interests of doctrinal conservatism: "Their success in history is the counterpart of the failure of the opposite view to commend themselves—to hold their ground in battle. They represent the 'survival of the fittest' in doctrine under the severest possible strain. . . . Yet men fling it aside as if this simple fact that it is old—has survived all this brunt of battle—were sufficient without further ado to condemn it! It is not explained why in every other sphere, the surviving product in an evolutionary process should be the fittest, and dogma alone should be an exception." Orr's capacity to marshal evolution in the cause of orthodoxy sprang from his insistence that "genuine evolution illustrates a law of continuity." Evolution, he explained, "is not a violent breaking with preceding forms, but proves its legitimacy by its capability of fitting into a development already, perhaps, in large measure accomplished."[76]

There remained implacably dissenting voices, of course. Duns, for example, never relented, and some Free Church highlanders, who were dismayed at the relaxing of subscription standards after the Declaratory Act of 1892 and were behind the overtures against Drummond a couple of years later, remained profoundly opposed to Darwinism in every shape and form.[77] But, taken in the round, evolutionary theory was pretty thoroughly domesticated to the intellectual leadership of the Scottish Presbyterian *mentalité*. After all, as Rainy insisted in April 1884 during the University of Edinburgh's tercentenary celebrations, there was every desire that "the Church's mind . . . should be in living sympathy with the march of science."[78] Later, when reflecting with his students on his earlier days, Iverach recalled how "glad for one" he was that "the Church did not condemn the law of gravitation"; it had wisely delivered no judgment "about geology and chronology"; nor had it "ever condemned evolution." "On such questions as these" he mused, "the Church should keep an open mind."[79]

Calvinism in Crisis

Compared with similar communities in other locations, Scots Presbyterians accommodated evolution with relative ease. But their accommodation must be understood in the context of other issues that traumatized Scotland's Calvinist culture during the final decades of the nineteenth century. Darwinism was never outlawed in the ecclesiastical courts of Scotland, as Iverach was delighted to report, but the same toleration was not extended to other matters. At the meetings of the Free Church General Assembly in Edinburgh during the late 1870s and early 1880s, a furor blew up over the new biblical criticism, resulting in a public spectacle. The whole affair congregated around the deeply disturbing theories of William Robertson Smith, the Free Church's professor of Hebrew in Aberdeen and, later, arguably Britain's most brilliant Orientalist. Compared with Smith's far-reaching proposals—which reworked the compositional history of the Hebrew scriptures, fundamentally revised the traditional understanding of biblical authorship, and uncovered an alarming archaeo-anthropology of Semitic sacrificial ritual—the challenge of Darwinian evolution was far from monumental. Revisit-

ing that whole controversy exposes the relative *unimportance* of Darwinian biology to the theological interests of Scottish Calvinists at the time. For when set
alongside the seismic provocations that Smith heaped on their traditional culture, the Darwinian issue simply receded into the background.

Smith's fertile intellect, rather ironically as it turns out, was profoundly animated by an evolutionary imagination, albeit not of the more narrowly Darwinian
variety. The shape that this impulse assumed in Smith's orientalism attracts our
attention not only because it exposes the far-reaching theological consequences
of his historicist methodology but also because it expands our awareness of Scottish Calvinism's engagement with evolutionism. Smith's mobilization of evolutionary modes of analysis in the context of higher criticism and the anthropology
of Semitic ceremonial merit reexamination in their own right, of course. But they
are important here because what were taken to be their shocking revelations shift
the balance of evolution's effect on Calvinist Scotland *away* from Darwinian natural
selection, species change, and heritable variation and *toward* nonbiological evolutionary forces internal to textual scholarship, conjectural prehistory, and primitive
cultural anthropology.

Born in 1846, William Robertson Smith disclosed precocious abilities in science, mathematics, and languages early on and at fifteen entered the university
of Aberdeen as a student.[80] So sparklingly brilliant was Smith that even Alexander Bain, whose intellectual constitution was irrepressibly antagonistic to a
Free Church outlook, considered him the most outstanding student he had ever
taught. By the early 1870s Smith had already published a range of scientific papers on geometrical reasoning, the fluxional calculus, and the flow of electricity.[81]
So impressive was his scholarly reputation that he was appointed to the chair
of Hebrew and Old Testament at Aberdeen's Free Church College at the age of
twenty-four.[82]

It was his 1875 entry on the Bible for the *Encyclopaedia Britannica* that first
thrust Smith into the public arena; here he showed his enthusiasm for the textual
criticism emanating from Germany.[83] Basic to his account was the fundamental
distinction he drew between the priestly and prophetic traditions in ancient Israel, the former ordinarily finding the "progressive ideas" of the latter "distasteful
to their natural conservatism and aristocratic instincts." His method of analysis
had implications for the dating of various Old Testament books, and he reviewed
the evidence for assigning to the Deuteronomic code a date that placed it in the
much later prophetic era. This meant that "the traditional conception" mistaken
for the "present shape of the Pentateuch" was actually "subsequent to the occu-
pation" of Canaan. The Pentateuchal legislation, then, was just simply "not one

narrative carried on from age to age by successive additions, but a fusion of several narratives which partly covered the same ground and were combined into unity by an editor."[84]

At first, only ripples of irritation were discernible. Archibald Hamilton Charteris, professor of biblical criticism at the University of Edinburgh, anonymously expressed concern in the *Edinburgh Courant* that Smith's article had appeared in "a publication which will be admitted without suspicion into many a household."[85] Others soon seized upon the matter and brought charges of heresy in 1878.[86] A litany of concerns soon surfaced: the date he assigned to the Levitical laws, the doubts he threw on the accuracy of particular texts, his understanding of the nature of prophetic prediction, and his detection of fictional material in certain biblical books.[87]

Proceedings were presently instituted and thus began Smith's protracted interrogation. Encouragement from such eminent continental scholars as Abraham Keunen, Ludwig Diestel, Paul de Lagarde, and Albrecht Ritschl, offered during the long, drawn-out hearings, may have meant much personally to Smith, but in the minds of his enemies, commendation from such quarters only justified their condemnation. Had they known about it, they would have thought the same about the support Smith received from the impish Richard Burton, with whom he had travelled in the Arabian Hijaz and who quipped, in evident delight: "What the Devil (a Ruskinism, there is no such body) will the Assembly say after the merry jig you have executed upon their pet corns? Dear, dear! So Moses did not write the books of Moses! (As if anybody ever believed he did.) If you republish, read (unless you have read) Spinoza, who proves the later date philologically."[88]

Smith was acquitted of the charges, but the issue did not go away. Far from it. His application of the Kantian categories to the record of divine revelation, for example, continued to disturb. Because he thought divine revelation analogous to the noumenon, it was necessarily only apprehended through "a series of subjective impressions" and consequently the personality of a prophet was inevitably impressed on his prophetic utterances. The prophet, Smith affirmed, was no "mere lyre struck by the plectrum of the spirit." To the contrary, all revelations presented to the mind were "deeply coloured" by the prophet's social and ethical *Zeitgeist*.[89]

And then there was the worrying direction his research took when he read the work of John Ferguson McLennan, the Scottish lawyer and parliamentary draftsman. Back in 1869–1870, in a three-part article on "The Worship of Plants and Animals" for the *Fortnightly Review*, McLennan had put forward an audacious theory of totemism, according to which a natural object is adopted by a tribal group as an emblem of the clan.[90] Earlier still, in his *Primitive Marriage* (1865), he had painted a picture of early humans as savages living in a promiscuous horde and put forward an extraordinary account of the matriarchal and poly-andric origins of civilization ultimately rooted in the unintended consequences of female infanticide.[91] These narratives were grounded in McLennan's social evolutionary outlook and relied heavily on a supposition propagated with much ardor by Victorian evolutionary anthropologists, namely, the idea of survivals.[92] It was an explanatory device that drew attention to the persistence of functionless—and often superstitious—traits in societies that lingered from earlier phases of social evolution.[93] Now Smith applied this entire line of thinking to the Bible in an essay for the *Journal of Philology* on animal worship in the Old Testament, in which he explored such related themes as animal gods, totem tribes, food prohi-bitions, exogamy, matriarchal kinship, levirate marriage, and laws of incest.[94] These researches drove him to the conclusion that his own findings were "re-markably confirmatory of Mr Mclennan's theory—a theory framed almost ab-solutely without reference to the Semitic races, but which nevertheless will be found to explain the true connection of a great number of facts which have hith-erto remained unexplained and almost unobserved."[95]

If Smith thought that this imaginative reworking of Semitic mores would lie buried in the dusty tomes of academia, he was mistaken. On Thursday, 29 July 1880, the Scottish newspapers printed a statement by Rev. George Macaulay, minister of the Roxburgh Free Church in Edinburgh, informing readers of his resolve to press the Church authorities "to issue an edict peremptorily prohibit-ing Professor Smith from the exercise of his functions as minister and professor in the Church." The denomination, he insisted, was in grave peril, and its tradi-tional standards had to be protected from Smith's "pernicious views."[96] Smith's latest article had precipitated this attack, and Macaulay appended a statement to the resolution fleshing out the major causes of concern. Smith's account of ancient Semitic marriage codes and animal worship were "so gross and so fitted to pollute the moral sentiments of the community" that they could not be aired beyond "the closed doors of any court of this Church." His reckless speculations would serve only "to destroy all reverence for God and for his Holy Word."[97] These allegations were the opening shots in the last phase of the five-year cam-paign to unseat Smith. Within a few short months, in the early hours of 25 May 1881, the General Assembly of the Free Church declared "that Smith's occupancy of his chair was no longer safe or advantageous for the Church."[98]

Though his status as a minister of the Free Church was not affected by this expulsion, Smith refused to accept a salary from that position on the principle that he could never consent to eat the bread of a church he was not permitted to serve.[99] Subsequently, he assumed general editorship of the *Encyclopaedia Britannica*, declined several approaches from Harvard, held the post of librarian at Cambridge, and, eventually, in 1889, just five years before his untimely death at the age of forty-seven, succeeded to the Thomas Adams Chair of Arabic at Cambridge. The libel case, Charles Raven always maintained, "broke his career and shortened his life."[100] In his last days, as Smith drifted in and out of consciousness, he would reportedly surface from time to time and shout in pained agitation, "Moderator, Moderator," as though reliving the trauma of the heresy trial.[101]

Despite Smith's extraordinarily brilliant defenses, remarkably fluid speech, and sharp wit, there was no place for his scholarship in the Free Church. Certainly he had allies, but the currents of Scottish Calvinist culture were flowing too strongly in a different direction. Nonetheless, as Smith left the Edinburgh Assembly Hall for the last time, his departure did not signal any rejection of the longer tradition in which he had been raised. At least that is what he said. He remained a Free Churchman all his life, believing to the end that he had neither said nor written anything that went against the spirit of the Westminster Confession of Faith.[102] His opponents did not see it that way. In their eyes, he had committed the most flagrant acts of treachery. To Dr. Robert Young, who anonymously put out the pamphlet "Infidelity in the Aberdeen Free Church," Smith had done nothing less than comprehensively undermine the morality, authority, and integrity of scripture. Even his friend Dr. James Candlish lamented the "dangerous and unsettling tendency" of Smith's outlook.[103] Later, more personal attacks came. The poet and literary critic William Ernest Henley, writing in the *Scots Observer* in 1889, sniped that even though "all Scotland held him in flattering respect, or still more flattering horror" Smith was more likely "part Voltaire without wit, and part Spurgeon without eloquence."[104] For his own part, Smith was not best pleased. Robert Rainy had finally sided with Smith's opponents, and his recommendation had been decisive. Smith never forgot that. In years to come, he reportedly advised a prospective ordinand never to trust Rainy—"he's a Jesuit."[105] To his friend J. S. Black, he was yet more direct: "I don't think that slimy cold-blooded reptile Rainy will stop till he has got the whole Church into a hole from which it can't get out again. He must be assassinated."[106]

The public spectacle of the Robertson Smith case and the lingering odor that for decades attached itself to Free Church culture cannot be ignored in any account of Scottish Calvinist dealings with Darwin. Of course, the heritage of the Scottish Enlightenment, with its enthusiasm for scientific pursuits, predisposed Presbyterians to engage positively with the achievements of modern science. But

Smith's profoundly disturbing interventions, which struck terror to the heart of Calvinist identity by unsettling the sacred foundations on which its heritage was erected, was particularly significant. His dazzling scholarship was infinitely more threatening than Darwin's claims about descent with modification, species transmutation, and the operations of natural selection. Apart from anything else, if energy was to be expended in rooting out modern skepticism and academic infidelity—the forces of enlightened darkness—Darwinian evolution paled into insignificance beside another target, namely, the subversion of scripture and the naturalization of Judeo-Christian praxis that Smith had evidently embraced. Thus, while some ambivalence to Darwin is clearly detectable during the 1860s and early 1870s, prior to the Smith debacle, by the late 1870s, 1880s and 1890s, the statements of intellectual leaders like Flint, Matheson, Iverach, Calderwood, and Drummond demonstrate the extent to which evolution had penetrated the thought-forms of the Scots Presbyterian intelligentsia. Indeed, something of the lengthy shadow that Smith's legacy threw over Darwin's fortunes in Calvinist Scotland may be gleaned from a quip that appeared in the *Edinburgh Echo* when Drummond's *Ascent of Man* was published in 1894. "One cannot take up this recent work without imagining the storm of holy wrath which would have been stirred in the Free Kirk a quarter of a century ago if the 'Ascent of Man' had then been published," the writer declared. Evidently, he continued, "Professor Robertson Smith has not lived in vain."[107] Another writer in the *Aberdeen Free Press* concurred. The presence of "a teacher of Darwinism" in the Free Church College might have been thought worthy of remark. "Yet so changed are the times in twenty years that, while a storm overwhelmed Professor Robertson Smith," he went on, "there is now scarce a ripple upon the ecclesiastical water."[108] What is ironic, of course, is that the most profound engagement with evolution in the Scottish Calvinist tradition was actually to be found in Robertson Smith himself as he mobilized its motifs for causes far more radical than any of the Darwinian peacemakers we have so far encountered.

Canonical Evolution and Mimetic Cannibalism

Ever since his student days Smith had repudiated Spencerian evolution and was sure that Spencer was mistaken to base his account on "the correlation of physical forces." Later, when they met, Smith reportedly found him to be rather "tedious."[109] And he was no more enamored of what he took to be John Tyndall's materialism. In fact, Smith had been in Belfast in August 1874 and witnessed firsthand the infamous presidential address that Tyndall, the prominent physicist and friend of Darwin and Huxley, had delivered to the British Association for the

Advancement of Science.[110] The young twenty-eight-year old Smith had come, hotfoot from a summer in continental Europe, to keep himself abreast of the latest scientific fashions. No doubt he was interested to see how Tyndall would acquit himself, given his deteriorating relationship with Smith's friend and teacher, the renowned Edinburgh natural philosopher, Peter Guthrie Tait, one of the fiercely brilliant "wild men" of Victorian natural philosophy.[111] Smith was comprehensively unimpressed. He frankly thought Tyndall's lecture "wretched" and decided that he would "pitch into it" in the pages of the *Northern Whig*,[112] where he did not pull his punches. Tyndall's grasp of the Middle Ages, Smith told his readers, was "at least a century behind the present state of scholarship," and his knowledge of the early development of atomism nothing short of pitiful.[113] To crown it all, Smith gleefully pointed out that Tyndall had uncritically relied on the dubious authority of John William Draper, whose *History of the Conflict between Religion and Science* appeared earlier that year in the same series as Tyndall's *The Forms of Water.*

In thus taking Tyndall to task in the public press, Smith gave every impression of standing alongside defiant opponents of evolutionary theory. Nothing could have been further from the truth. There is good reason to suppose that he absorbed more expansively evolutionary values than any of his colleagues in the Free Church tradition who recorded their support for the new natural history. As Keith W. Whitelam observes, "The notion of evolutionary progress runs as a dominant thread throughout his writings colouring his attitude to religion, society, and history . . . in short, it underpins virtually every aspect of his thought."[114] That catholic evolutionism now takes center stage.

Several years before he crossed swords with Tyndall in Belfast, Smith had already advertised his enthusiasm for an evolutionary reading of biblical history when he published his thoughts on "The Question of Prophecy in the Critical Schools of the Continent" in 1870. Here he urged that a "tradition that violates the continuity of historical evolution and stands in no necessary relation to the conditions of the preceding and following age must be untrue; and, above all, an ancient writing which is no frigid product of the school, but is instinct with true life, must be the product of that age which contained the conditions of the life it unconsciously reflects."[115]

These hints were just the beginning of a sustained evolutionary reading of the Bible's textual development. His controversial entry on the Bible for the *Encyclopaedia Britannica,* on which he embarked within a few months of his quarrel with Tyndall, for instance, was entirely cast in the language of evolutionary gradualism. The biblical books, he began, "set before us the gradual development of the religion of revelation," the Old Testament was a record of the "struggle and Progress of Spiritual Religion," and the working out of the tensions between "the spiritual faith" and polytheistic nature worship required the former "to show constant powers of newer development . . . proving itself fitter than any other belief to supply all the religious needs of the people." It was precisely in the context of this "struggle between spiritual and unspiritual religion" that the strains between the prophetic and priestly impulses in ancient Israel were to be located.[116] Smith's whole understanding of the history of the biblical canon, in fact, was bound up with an evolutionary appreciation of the progress of Israelite ceremonial as expressed in liturgical development. The record of the scriptural writers could thus only be grasped in the context of Israel's socio-ritual evolution, and this provided the hermeneutic key to disentangling the history of the compilation of the Hebrew canon.

All of this was indicative of Smith's conviction that the corpus of scripture was subject to the natural laws of textual evolution. After all, he was fully aware that language itself had undergone historical transformation. All Semitic dialects had sprung from an ur-tongue—an original language—and "each member of the group had an independent development from a stage prior to any existing language."[117] Linguistic, philological, and psychological evolution were intimately intertwined, for, as he put it in 1876, the "only idea of moral and spiritual evolution possible to us, is that of evolution in accordance with psychological laws."[118] Later, in his 1881 lectures, published as *The Old Testament in the Jewish Church*, he again directly linked the different phases of Hebrew legislation with Israel's developing social economy. As Gillian Bediako observes, "Smith set the progress of Old Testament religion in the context of human evolution."[119] This is not surprising. The principle of evolutionary change was so deeply ingrained in the documentary hypothesis of the higher critics that John Rogerson, excusing his lack of attention to the *Origin of Species* in his survey of nineteenth-century Old Testament scholarship, records that "one gets no indication of an almighty conflict occasioned by Darwin."[120]

If evolutionary motifs facilitated Smith's inclinations toward biblical criticism, they were no less crucial to his work on the genealogy of primitive sacrifice.[121] His 1887 Burnett Lectures, later published as *The Religion of the Semites*, are illustrative. In setting out his methodological wares in what has been described as "one of the founding texts of modern anthropology," Smith had repeated recourse to evolutionary diction.[122] He spoke of arrested development in the "evolution of Semitic society," of the changing connections between "religion and kinship" depending on "stage of society," and of the need to link "more advanced ideas" with "a higher stage of social development." Traditional religious practices had evolved slowly over the course of centuries and for that reason bore the stamp of "habits of thought characteristic of very diverse stages of man's intellectual and moral development." Ritual forms thus disclosed historical sequencing analogous to geological strata: "The record of the religious thought of mankind, as it is embodied in religious institutions, resembles the geological record of the history of the earth's crust; the new and the old are preserved side by side, or rather layer upon layer. The classification of ritual formations in their proper sequence is the first step towards their explanation, and that explanation itself must take the form, not of a speculative theory, but of a rational life-history."[123]

Clearly, whatever his reservations about Tyndall or Spencer, Smith emphasized the emergent properties of evolving religion and the progressive nature of divine revelation. Indeed, in a passage redolent with Darwinian vocabulary, he announced that the "communities of ancient civilisation were formed by the survival of the fittest, and they had all the self-confidence and elasticity that are engendered by success in the struggle for life." What facilitated Smith's evolutionary reading of religious history was his assurance that practice must take precedence over dogma and that the customary segregation of religious observance from ordinary life must be broken down. Because "antique religions had for the most part no creed" and "consisted entirely of institutions and practices," their study "must begin, not with myth, but with ritual and traditional usage."[124] The fundamental character of institutions and practices meant that, for Smith, the basic unit of analysis was the community, not the individual. Individuals did not choose their religion in ancient times; rather, religion came to them as an integral component of the social obligations laid upon them by virtue of their position in the family and clan.

Having thus laid down the conceptual tracks along which his subsequent analysis would run, Smith turned his attention to the evolution of Hebrew ritual. Drawing again on the inspiration of McLennan, he fastened on the central role of sacrifice in the production and reproduction of social solidarity. The ritual slaughter and consumption of a totemic victim ordinarily regarded as taboo was taken by Smith as an exercise in sacramental communion between the human and divine worlds and as the means of maintaining an aboriginal sense of clan union and tribal solidarity.[125] In Smith's telling, primitive society's identification with a totem victim meant that where "an animal is sacrificed, the sacrificer and the deity feast together, part of the victim going to each." This practice meant that in the observance of the sacrificial feast "the god and his worshippers are *commensals*."[126] The holding together of the primeval community incorporating tribe and deity was the primary element in sacrificial rites; the very act of ceremonial consumption was symbolic of the mutual social covenant between the clan and its god. Sacrifice was thus central to ancient tribal life, for "participation in the flesh of a sacrosanct victim, and the solemn mystery of its death" were rendered legitimate on the understanding that only in this way could "the sacred cement be procured which creates or keeps alive a living bond of union between the worshippers and their god."[127]

Perhaps the most grisly instance of primitive ritual communion was colorfully elaborated in Smith's account of the sacramental feast associated with Arabian camel sacrifice. Once the first wound had been inflicted, the tribal leader "in all haste drinks of the blood that gushes forth," and "the whole company falls on the victim with their swords, hacking off pieces of the quivering flesh and devouring them raw with . . . wild haste." To Smith, the "plain meaning" of this gruesome ritual was "that the victim was devoured before its life had left the still warm blood and flesh . . . and that thus in the most literal way all those who shared in the ceremony absorbed part of the victim's life into themselves." Such primordial gorging, moreover, was not restricted to animal sacrifice. Cases of human ritual slaughter and ceremonial cannibalism were certainly to be found, and Smith was sure that their progressive demise was *not* on account of any natural human revulsion against consuming human flesh. "What seems to us to be natural loathing," he observed, "often turns out . . . to be based on a religious *taboo,* and to have its origin not in feelings of contemptuous disgust but of reverential dread. Thus . . . the disappearance of cannibalism is due to reverence, not to disgust, and in the first instance men only refused to eat their kindred."[128] To Smith, then, prehistoric practices of sacrifice were bound up with a primeval belief that the bloody wolfing down of "the gobbets of throbbing flesh, newly-killed of their fellow tribesmen"—as George Elder Davie colorfully puts it[129]—provided the means of revitalizing an aboriginal sense of clan union.

As these sacrificial systems evolved, their surface features were stripped of such grotesque transactions. The "primitive crudity of the ceremonial was modified and the meaning of the act is therefore more or less disguised," Smith explained. Its origins were obscured, but its deep structure was not obliterated. The significance of the idea of survivals, so central to Victorian anthropology, clearly surfaced.[130] Thereby Smith explained the practice of a hurried consumption of the Passover feast as "having come down from a time when the living flesh was hastily devoured beside the altar before the sun rose." These and other points of contact between what Smith called "the most primitive superstition" and later Hebrew ceremonial regulations about ritual uncleanness, forbidden foods, and the like were evidence that certain rites "preserved with great accuracy the features of a sacrificial ritual of extreme antiquity." Smith also acknowledged the sense of affective ambivalence and emotional crisis that attended such performances when he elaborated on what he called the "close psychological connection between sensuality and cruelty" which surfaced "in ghastly fashion in the sterner aspects of Semitic heathenism." The reason was not hard to find: "the same sanctuaries which, in prosperous times, resounded with licentious mirth and carnal gaiety, were filled in times of distress with the cowardly lamentations of worshippers, who to save their own lives were ready to give up everything they held dear, even to the sacrifice of a firstborn or only child."[131]

All of this was profoundly at odds with the elemental convictions of traditional Calvinists in Scotland, not least when Smith's observations touched upon eucharistic motifs. His thinking on this subject was especially alarming when he spoke of "an older rite, in which the victim was not a mere effigy but a theanthropic sacrifice, *i.e.* an actual man or sacred animal, whose life . . . was an embodiment of the divine-human life." Again, as he reached toward a conclusion to his account of Semitic religion's fundamental institutions, he worryingly noted that "the various aspects in which atoning rites presented themselves to ancient worshippers have supplied a variety of religious images which passed into Christianity, and still have currency. Redemption, substitution, purification, atoning blood, the garment of righteousness, are all terms which in some sense go back to antique ritual."[132] Plainly, the whole economy of Judeo-Christian soteriology, though purged of primitive orgiastic excesses, was nonetheless rooted in some form of sublimated cannibalistic social memory.

It is not surprising that this remarkable account attracted the admiring attention of Sigmund Freud. Freud's enthusiasm for the "fascinating" and "admirable work" of the "clear-sighted and liberal-minded" Smith sprang from his psychoanalytic interest in taboo, in the relevance of totem sacrifice for the Oedipus complex, and in the profound sense of equivocation that attended a ritual in which the very thing most revered—divinity incarnate—was ceremonially slain. But he was convinced of the profundity of Smith's recital and devoted considerable space to sketching in its lineaments. Having completed his synopsis, Freud starkly drew out the relevance of Smith's account for psychoanalysis. The "totem animal is in reality a substitute for the father," he observed, "and this tallies with the contradictory fact that, though the killing of the animal is as a rule forbidden, yet its killing becomes a festive occasion—the fact that it is killed and yet mourned. The ambivalent emotional attitude, which to this day characterizes the father-complex in our children and which often persists into adult life, seems to extend to the totem animal in its capacity as substitute for the father." "Fantastic" as Freud himself admitted this thesis might sound, it showed to his own satisfaction how psychoanalysis could blend "Darwin's theories of the earliest state of human society" with anthropological work on totemism.[133]

For all that, Smith's tough-mindedness stood in marked contrast to Freud's own "ambivalence" in rebroadcasting the idea of cannibalistic nostalgia. Whereas Smith presented his account with an unadorned frankness—he evidently preferred to take his savagery neat—Freud was deeply conscious of its disturbing implications for modern high culture. Indeed, although he acknowledged his indebtedness to Smith, Freud found cannibalistic nostalgia just a bit too much. In its beginnings perhaps there were, after all, "no deeds, but only impulses and emotions." In the case of primitives, no less than contemporary neurotics, Freud speculated, "the mere hostile *impulse* against the father, the mere existence of a wishful *phantasy* of killing and devouring him, would have been enough to produce the moral reaction that created totemism and taboo. In this way we should avoid the necessity for deriving the origin of our cultural legacy, of which we justly feel so proud, from a hideous crime, revolting to all our feelings."[134]

Freudian enthusiasm notwithstanding, it is notable that Smith's startling and disturbing account of Semitic ritual was grounded in a thoroughly evolutionary understanding of religion as a developmental social formation. To be sure, Smith's evolutionism was much more conditioned by Enlightenment ideas of progress and Victorian conceptions of social evolution than by anything drawn explicitly from Darwin. But that only means that, in Smith's hands, non-Darwinian evolution was far more threatening to the traditional interests of Calvinist Scotland than Darwin's so-called "dangerous idea."[135]

◆ ◆ ◆

DESPITE SOME INITIAL EQUIVOCATION in their early dealings with Darwin, Scots Calvinists by and large rapidly made their peace with biological evolution. In most cases they had found, to their own satisfaction at least, mechanisms for preserving their foundational faith in teleology in the face of naturalistic Darwinism and even for advertising evolutionary natural history as a means of putting the idea of design on a firmer footing. Indeed, as the twentieth century dawned, it was clear that some were happily deploying evolutionary rhetoric to shore up their conviction about the psychosomatic unity of the human race, creedal history, and progressive revelation. These developments reflected the long-standing enthusiasm for scientific endeavor within that culture.

At the same time, the local commotion over William Robertson Smith presented a different set of challenges. However much he thought he was using modern scholarship to sustain, not to subvert, Christian theology, Smith struck terror to the hearts of traditional Calvinists. Disturbed to the center of their cultural being, they directed their collective energies to weeding out what they saw as Smith's importation of rank profanity from Germany, "that fountain of all poison," as Rev. Alex McCraw put it in a letter to the *Scotsman*.[136] This was the front on which battle needed to be engaged, and Darwin's biological theory seemed comparatively tame. If anything, it was Smith's mobilization of an older, non-Darwinian form of social evolution that animated his challenging proposals. Even theological advocates of biological transformism rarely perceived the implications of Smith's version of evolution for the textual reconstruction of scripture and the anthropological readings of religious rites that he so imaginatively elaborated. While Robert Rainy, for example, was sanguine about applying evolutionary principles to human origins, he was—as his biographer notes—"critical of scientific methods in religion."[137] Smith, by contrast, was engaged in nothing less than an expansion of the scope of evolution's empire to encompass the entire archaeo-anthropology of Judeo-Christianity. The fate of evolutionary theory among Scottish Calvinists, I suggest, was shaped in various ways by the legacy of Enlightenment science, the local contingencies of identity politics, the spectacle of public theater enacted on the floor of the Edinburgh New College Assembly Hall, and the struggle over the application of scientific methods to the historical study of Semitic scripture, sacrifice, and society.

CHAPTER THREE

Belfast, the Parliament of Science, and the Winter of Discontent

J osias Leslie Porter was deeply troubled. As he surveyed the intellectual landscape of his day, he could only discern "melancholy proofs that science and philosophy" were no longer "safe guides in the education of a people." What disturbed him most was the widely circulated "dogma that life is evolved from material atoms." In particular, the contemptible declaration of "Professor Huxley" that thoughts "are the mere expressions of molecular changes in . . . matter" was alarming. The idea that "animals are but conscious automata" was bad enough, but infinitely worse was the suggestion that human beings, in all their "thinking, and willing, and moving, and acting," operated "as mere machines, under the inevitable and irresistible impulse of external forces."[1]

As Professor J. L. Porter (1823–1889) formally opened the new academic year at the Presbyterian College in Belfast on Tuesday, 19 November 1874, he told colleagues and students alike that he stood ready "to show that not a single scientific fact has ever been established . . . from which these dogmas can be logically deduced." The "effect of such scientific and philosophic teaching upon the nation if unchecked and uncorrected" would be nothing short of calamitous. "Would it not be to quench every virtuous thought, to repress every noble aspiration, to extinguish purity and holiness and self-denying beneficence?" Theological colleges were more needed than ever before if the "evil which contaminates philosophy" was to be countered. For it was only in consecrated spaces like the General Assembly's seminary that philosophy and science could "be preserved from those wild theories and reckless speculations which make them noxious to mankind." Only in such settings could science be kept within its proper bounds. And careful policing was badly needed for, he insisted, "the deification of science was one of the main causes of the horrors of the French Revolution."[2]

Robert Rainy had delivered his inaugural address for the new year to the Edinburgh New College students the previous Wednesday. But there the rhetorical tone was notably different from Belfast, notwithstanding the two colleges' seriously similar confessional identities. Rainy received enthusiastic applause when he announced that it would "be absurd to draw an arbitrary line, and say that so much evolution, and no more, was to be admitted into thoughts of the history of things, and that so much, and no more, should be held to comport with the method of God." To be sure, Rainy was fully aware that there was "a significant connection between such questions and atheistic tendencies," but the applause was renewed, the *Scotsman* reported, when he cautioned "his hearers against regarding men with suspicion simply because they might think . . . that evidence tended to establish the assertion that species originated by evolution." Science was entirely within its rights, he remarked, "in refusing to be regulated by what would prove for its purposes matters of doubtful disputation."[3]

The different, and conspicuously more aggressive, rhetoric that Porter adopted in Belfast cannot be attributed to the personal predilections of a scientific malcontent. For Porter *was* enthusiastic about scientific enterprises. He had just contributed a paper to the Geographical Section of the British Association for the Advancement of Science, which had convened a month or two earlier in Belfast, reporting on the scientific dimensions of his "Recent Journey East of the Jordan."[4] Besides, he was far from alone in the sentiments he voiced. When Henry Wallace (1801–1887), professor of ethics at Assembly's College, spoke to his students that same winter on the subject "Teachings of the British Association," he took exactly the same line. The recent Belfast meeting of the British Association had definitely got Wallace's attention. The dissemination of the "atheist principle," he told his students, "was manifestly the main aim of the president's address," and the findings he presented were not "dwelt upon so much for their scientific value, nor as records of progress, but merely to serve the cause of atheism." Scientifically, he continued, "Professor Tyndal's [*sic*] reasonings and inference from the real facts of science are as false as they are shallow," and his "experimental philosophy casts not one ray of light upon the origination of matter, or of life, or of mind, and none upon the destiny of either." Morally, it was even worse. Tyndall-style science, Wallace declared, echoing Porter's inaugural address, could "only sap the foundations of moral order, abolish the distinctions of virtue and vice, and 'guide' again to the atrocities of the French Revolution."[5]

That Wallace should choose the British Association as the subject matter for one of his ethics lectures points to the prominent place its recent Belfast meeting occupied in the imagination of the college's faculty during the winter of 1874. And it did so for good reason. What transpired at that event, the expressive style adopted by interlocutors, and the notoriety that the meeting acquired far and wide set the terms of the debate about Darwin in Calvinist Belfast for at least a generation. What could be *said* about evolution and what could be *heard* about it were shaped by the memory of the intellectual bombshell that John Tyndall, the 1874 president of the British Association, had detonated in the Ulster Hall during that eventful week in August. Local Presbyterian congregations the following Sunday heard about little else. And when Thomas Henry Huxley acerbically dismissed as "pigmies in intellect" the sermonizers who had hit back at Tyndall, he only fanned the flames.[6]

Rooted in the Scots settlement of the north of Ireland in the early years of the seventeenth century, Ulster Presbyterianism retained strong cultural links with its Scottish heritage.[7] In the early days, most of its clergy were educated at Glasgow University, where they were exposed to the moral philosophy of such Scottish Enlightenment luminaries as Adam Smith, Thomas Reid, and the Irish-born Francis Hutcheson. Mental science would thus become as central to Ulster Presbyterian intellectual life as it was to its Scottish counterpart and was likewise supplemented by a taste for natural philosophy and its practical applications in a developing industrial economy. In Belfast a host of improving societies—a Reading Society, a Natural History Society, a Literary Society, and the like—came into being in the decades around 1800. At the Belfast Academical Institution, which opened in 1814 to provide local training for the ministry, clerical candidates could take classes in mathematics, geography, and surveying from James Thomson (the father of Lord Kelvin) and in anatomy, physiology, and natural history from James L. Drummond.[8] Here too, as in Glasgow, the Scottish Common Sense Philosophy was instilled in students—a philosophy that urged that attention be directed to how minds actually work (rather than to theoretical prescription), and thus the so-called Baconian method of induction was reinforced. All of this conspired to stimulate a profound admiration, indeed enthusiasm, for the scientific enterprise and a deep-seated antagonism to the skepticism of Hume and the idealism of Berkeley.

This stance was further buttressed when the Queen's College opened its doors in Belfast in 1849, as the first two professors of philosophy—Robert Blakey and James McCosh—were devoted advocates of the Scottish Common Sense School.[9] Indeed, when he delivered his inaugural lecture in 1852, McCosh resorted to Abraham Cowley's 1663 paean "To the Royal Society" to advertise his personal devotion to the Baconian ideal:

> Bacon, like Moses, led us forth at last,
> The barren wilderness he passed,
> Did on the very border stand

Of the blessed promised land,

And from the mountain's top of his exalted wit

   Saw it himself, and shewed us it.[10]

In the field of mental science, McCosh stood foursquare behind his mentor Thomas Reid. Like Reid, McCosh was convinced that the inductive method was the only sure and certain way to proceed. The painstaking, slow, incremental gathering of data was much to be preferred to a priori theorizing. To understand the workings of the mind, it was no good to meditate on works of high theory; examining the concrete, everyday operations of common wits was the only way to make real progress. With a high reputation for his first book, *The Method of the Divine Government* (1850), McCosh further consolidated his interests in natural philosophy by contributing scientific papers to different meetings of the British Association on botanical subjects and collaborating with the Queen's College professor of natural history George Dickie.[11]

The Scottish intellectual tradition, it is clear, had delivered to Ulster Calvinists both philosophical and theological resources to foster the cultivation of a scientific culture in the north of Ireland during the first half of the nineteenth century. This aspiration was supplemented by the political circumstances in Ireland, where fears of Catholic domination fueled a rhetoric that portrayed the Church of Rome as deeply antagonistic to intellectual improvement, civil and religious liberty, and modern methods of acquiring sound knowledge. The memory of Galileo's fate at the hands of the Inquisition served to confirm in the minds of Ulster Presbyterians the fundamentally Protestant character of true science. As William Gibson, the Belfast minister and later professor at Assembly's College, made clear in his attack on the government's proposal to enlarge the endowment for the Catholic Saint Patrick's College in Maynooth, Rome's record had shown it to be "adverse to all truth, whether of science or religion."[12]

By the middle of the nineteenth century, then, the robust scientific culture that had been established in Ulster enjoyed the benediction of its Calvinist custodians. Such commitments, of course, were not incompatible with the strongly literalist reading of the Genesis narrative still enshrined in the culture of Assembly's College. When James G. Murphy, the professor of Hebrew, produced a pamphlet of a lecture on *Science and Religion before the Flood* in 1857, his lauding of the virtues of science was couched entirely within the assumption of the literal truth of the Mosaic narrative. For him, the beginnings of botany, zoology, and philology were all to be found in the Genesis account of Adam in the Garden of Eden. With these convictions, it is understandable that he was allergic to what he considered the facile progressivism that infatuated the nineteenth century. The idea of stadial social development from savagery, via pastoralism and agriculture, to civilization "savours too much of the complacency . . . of the sixth millennary of the human race," he observed. In fact, the so-called "savage" often evinced nobility and heroism, and the pastoral was often characterized by "the gentle and the contemplative." Developmental schemas thus reversed reality for, as he put it, "Civilisation marks the origin, and barbarism the middle ages of the human race."[13] Plainly, Murphy's portrayal of the nobility of scientific inquiry did not extend to the speculations of the nascent social sciences on the early human condition.

What, then, of Darwinism? How was his theory of evolution read in Belfast in the years before Tyndall's blast? And what long-term effect did the local hosting of the British Association exert on how Ulster-Scots Presbyterians engaged with the Darwinian dilemma? To get a handle on these questions, we need to turn first to the situation in the years between the publication of the *Origin of Species* in 1859 and the 1874 Belfast meeting. As we do so, we should remain attuned to the vital importance of rhetorical nuance in the whole affair, to the tone interlocutors adopted as they engaged in dialogue, to how audiences heard their pronouncements, and to how different spaces conditioned the forms of expression that speakers assumed.

The Lull before the Storm

During 1859 and the year or two following, concerns over science and religion in Belfast certainly did not center on Darwin's *Origin of Species*. Public debate on such matters was far more likely to revolve around a local phenomenon then gripping Ulster society. In 1859, the north of Ireland had found itself swept by a tide of religious revival—a spiritual awakening accompanied in part by "convulsions, swoons, visions, and other extraordinary effects that seemed to spread through the province of Ulster like an 'epidemic.'"[14] In medical treatise and popular print alike, this outburst of religious fervor was subjected to interrogation by sympathizers and critics. William MacIlwaine, the rector of Saint George's Church in Belfast's city center, welcomed the spiritual exercises that transformed personal life and fostered family devotions. But he was bitterly opposed to what he described as its physiological "excesses." For readers of the local press and the *Journal of Mental Science*, MacIlwaine conducted his campaign against "fanaticism of the wildest type," a fanaticism, moreover, that was class- and gender-biased toward "the ignorant, uneducated, hard-worked, and easily impressed class, and, in the proportion of nineteen out of every twenty, young and excitable females."[15]

Although James McCosh was no less troubled by its "physiological accidents," he found himself cast as the revival's chief apologist. His strategy was to relieve the movement of the burden of prejudice that its accompanying bodily manifestations had fostered in the minds of unsympathetic critics. To him these physical displays were indeed mere "accidents," explicable as purely natural reactions to extreme emotional experience and largely attributable to Irish ethnic temperament. But—and this is crucial—McCosh cast his diagnosis in the language of defense by insisting that the "deep mental feeling" that induced somatic convulsions was itself "a work of God."[16] For this strategy, his commitment to Scottish Common Sense philosophy stood him in good stead. His inductive cataloging of psychosomatic data and his finding a scientific explanation for the observed behavioral traits allowed him to prosecute a critical distinction between genuine inner religious experience and contingent external physical expression.

Evidently, if scientific scrutiny was to be brought to bear on religious sensibilities around this time, it was the 1859 religious awakening that most pointedly engaged interlocutors. So, whether on account of the revival and its pastoral legacy, or the storm waves generated by the publication the following year of *Essays and Reviews*, or because no threat of epic proportions was perceived in the early reading of *The Origin* itself—whatever the reason—the initial response in Belfast to Darwin's own proposals is conspicuous only by its relative absence. Where encounters were registered, they were markedly muted. Some isolated comments were made, but these were by no means universally hostile. The recently ordained George Macloskie, for example, made some tangential observations on the Darwinian theory in "The Natural History of Man" in 1862. While expressing some hesitancy about the universal claims of the Darwinian system, Macloskie paused to confirm that the principle of natural selection could account for human racial differentiation and provide good grounds for rejecting polygenism in favor of monogenism. Indeed, he insisted that he himself had "already employed this principle, to explain the diversity that exists in the different tribes of mankind, whilst the specific unity is still preserved."[17] To be sure, the deductive character of the Darwinian theory was, at this stage, bothersome to Macloskie, though in years to come—after he followed his teacher McCosh to Princeton in 1875—he would himself defend the necessity of scientific speculation and emerge as a major exponent of evolution among Old School Presbyterians in the United States.[18] At the time, though, he was still cautious. Darwin's theory, he surmised, outran the evidence and was "not confirmed by a sufficient induction of facts." He was troubled by the absence of persuasive intermediate forms and Darwin's inclination to play fast and loose with the standard definition of species.

Besides, while conceding that the "theory may be beautiful," he was fully aware that it was being regarded "as a covert thrust against the testimony of scripture."[19] Still, Macloskie's piece was conspicuously bereft of the invective that would characterize debates in the aftermath of Tyndall's Belfast address.

In November 1863, Rev. William Todd Martin (1837–1915) provided his assessment, *Our Church in its Relation to Progressive Thought*, when he spoke to the Newry Presbyterian Young Men's Society. He had recently been installed as the local minister and later, after service in Newtownards, succeeded Henry Wallace as professor of ethics at Assembly's College in 1887. That year, as we shall see, he published a three-hundred-page evaluation of Herbert Spencer's cosmic evolutionism. The main thrust of his 1863 homily, however, was to express concern at the inroads being made by *Essays and Reviews* and the Pentateuchal criticism of John William Colenso, the first Anglican bishop of Natal, and Principal John Tulloch of the Established Church of Scotland, whose concessions to modern science and textual criticism, Martin believed, subverted traditional biblical authority. Such activities, he forecast, would "make shipwreck of their faith." He did pause to lament that the "Darwinian theory of development, and Sir Charles Lyell's theory of the Antiquity of Man, are conclusive proof that the tendency of philosophic thought is flowing away from evangelical truth, and towards scepticism."[20] But it was only later, in the aftermath of the Tyndall shock and in the wake of Herbert Spencer's cosmic philosophizing, that he focused more specifically on the evolution question. For the time being, the implications of the Darwinian intervention were scarcely perceived, thereby bearing out the suspicion that, on the religious front at least, *The Origin* was "upstaged"—to use Jim Moore's word— by *Essays and Reviews*.[21]

And then, during his inaugural lecture on "The Ministerial Curriculum" at Assembly's College in 1866, Robert Watts paused to make some remarks on Darwinism's irreconcilability with classical Paleyan natural theology. His purpose was to confirm the need for an educated ministry and also to undermine the naturalistic restrictions the Darwinian theory seemed to impose on traditional conceptions of causality. What troubled him was the idea that organic modification—in the hands of Huxley in particular—was proposed as operating "independent of any *ab extra*, controlling or directing cause—a development absolutely free from the guidance or interference in any shape, of any intelligence, whether Aeon or Demiurge." At *this* stage Watts's treatment of the subject was muted, cursory, and focused on the implications of Darwinian evolution for teleology. It was, as he put it, "simply to show that this theory has been constructed in violation of unquestionable primary principles . . . It is manifestly in conflict with the principle of causality."[22] For Watts, Darwin's proximate causes were no substitute for Paley's ultimate cause.

Doubtless other early passing evaluations of Darwin's theory lie buried in the archives of Belfast's intellectual past. But by far the most considered and expansive appraisal of the new biology was made not by a Presbyterian but by a businessman who was secretary to the Church of Ireland's Diocesan Synod and Council, Joseph John Murphy. Murphy was a leading light in a variety of Belfast scientific and literary circles, twice occupying the presidency of the Belfast Natural History and Philosophical Society.[23] In his 1866 presidential address, he subjected Darwin's theory to detailed scrutiny. Here he expressed substantial agreement with Darwin but demurred on the *all*-sufficiency of natural selection to account for organic variation in toto and in particular to provide a compelling explanation for the emergence of the eye.[24] In fact, Murphy's latter query was to achieve considerable prominence. Darwin himself cited it in *The Variation of Animals and Plants under Domestication*,[25] and Sir Arthur Keith excerpted what he called this "just criticism" in his Huxley lecture for 1923 and in his presidential address at the Leeds meeting of the British Association in 1927.[26]

For Murphy, of course, querying natural selection's ubiquity did not mean denying its activity, and he deployed the Darwinian model in his paper "The Origin of Organs of Flight," presented to the Field Club in 1869.[27] That same year he brought out the first edition of his two-volume *Habit and Intelligence*. Its title revealed the work's fundamental orientation—illustrating the role of habit and intelligence in organic modification. Throughout, Murphy displayed a Lamarckian-sounding enthusiasm for the role of what he called "self-adaptation," in conjunction with natural selection, in the history of life. But these twin mechanisms were insufficient by themselves to explain organic complexity without the role of an "Organizing Intelligence." That the law of natural selection had been "proved," Murphy had no doubt; that "all organic species have been descended from one or a few germs" he likewise had no cause to deny. His only reservations were that the Darwinians were prone to claim near omnipotence for the natural selection mechanism and that recent geophysical research on the age of the earth required a more rapid evolutionary tempo than Darwin was ready to admit.[28] Reservations notwithstanding, Murphy later insisted that in *Habit and Intelligence* he had gone "about three-quarters of the way" with Darwin. But within a few years, he confessed, the three-quarters had been reduced to half.[29] And the figure continued to fall. His audience at the November 1872 meeting of the Belfast Natural History and Philosophical Society learned from their president that Kelvin's figures for the age of the Earth presented "a great difficulty in the way of Darwin's theory of the origin of species." "Eternity would not be long enough" for the Darwinian mechanism to derive all living beings "from the simplest animalcules."[30] By the time his next presidential address came around in 1873, Murphy was vigorously contending for an anti-Darwinian account of evolution. He could now only concur with Darwin on the *fact* of evolution. Other than that, he could not "now agree with the distinctive parts of Darwin's theory at all." What Murphy had come to enthusiastically endorse was a *non*-Darwinian, theisized version of evolution "guided by Intelligence."[31]

Yet for all that, prior to 1874, there is little evidence to suggest that evolutionary theory per se was causing any profound anxiety in Calvinist Belfast. To be sure, from time to time commentators expressed reservations—particularly over the ubiquity of natural selection. But these lacked vitriol or vituperation. Things dramatically changed during the winter months of 1874–1875, however, as a concerted assault was orchestrated against the new biology. A massive departure from the seeming complacency of the previous fifteen years or so was now underway.

The British Ass

The Belfast newspaper, the *Northern Whig*, enthusiastically announced the arrival of the "Parliament of Science"—affectionately known as the British Ass—on Wednesday, 19 August 1874.[32] It was "no ordinary honour and privilege," its readers were told, "that Belfast should enjoy this year a renewal of the visit paid . . . two-and-twenty years ago by the philosophers of the British Association." The meeting, rather ironically, was welcomed to the city as a temporary respite from "spinning and weaving, and Orange riots, and ecclesiastical squabbles." Nevertheless, "some hot discussions" were predicted "in the biological section" between advocates of human evolution and those "intellectual people—not to speak of religious people at all—who believe there is a gulf between man and gorilla." The editor was convinced that the "Irish people . . . will give a patient hearing to the philosophers notwithstanding." The newspaper welcomed the association to the city "with all heartiness," and afterward credited the resolution of a local workers' strike to its presence.[33] Even after Tyndall's dramatics, the Presbyterian *Witness* reflected on how "heartily glad" it also was to welcome the association to the city. In what was doubtless a jibe at the Catholic community, it remarked that three hundred years earlier it would have been different, for that was a time when "the advancement of science" was "not much in favour with the Church." Still, while happily reporting that the days of religious antagonism to science were a thing of the past, the *Witness* coolly remarked: "Would that we could add that scientific men are always animated by an equally friendly spirit towards true re-ligion!"[34] There was a definite frostiness in the air.

The Belfast meeting was to be what James Moore and Adrian Desmond have called "an X Club jamboree," with its priestly coterie of Huxley, Joseph Dalton Hooker, John Lubbock, and Tyndall, all speechifying.[35] This "club," an informal fellowship of like-minded scientists, had been meeting together for dinner for nigh on a decade and included key members of the rising scientific establish-ment. All were committed to the emancipation of science from established civil and theological authorities, and they used their collective cultural clout to ad-vance naturalistic thinking at every opportunity. If an assault was to be mounted by the new scientific priesthood on the old clerical guardians of revelation and respectability, scripture and social status, then what better venue could there be for a call to arms than the British Association meeting in Calvinist Belfast?

Tyndall, a native of County Carlow, was only too happy to favor a militant strategy, and his pugnacious reputation did not escape the notice of his native audience. A few hours before his evening speech, readers of the *Northern Whig* were introduced to the president-elect as "a perfect Irishman in his controversies, flinging himself readily into the arena, and carrying on his wars with a thor-oughly Celtic vehemence and good humour."[36] He was determined not to disap-point. It was precisely because Tyndall felt that Belfast audiences did not want to hear him that he resolved to be even "less tender" in his speech than he ordinarily would. He might well have wished he had not let Huxley talk him into the whole business; but he decided not to listen to Huxley's advice to be "wise and prudent." As he began, it was reported, his voice was "rather jerky and uneven," but he evi-dently settled into things quickly enough and became "more impressive."[37]

Tyndall's performance did not fall short of expectations. In an Ulster Hall garnished with accompanying orchestra, Tyndall delivered—with nothing short of evangelical fervor—a missionary call to liberate science from theological con-trol. He began by presenting a thumbnail sketch of the history of the atomic theory since the time of Democritus, Epicurus, and Lucretius, then charted the triumph of Darwinian evolution in deriving "man in his totality from the inter-action of organism and environment through countless ages past." All of this was intended to show once and for all how science had progressively routed the forces of metaphysical dogma. The implications were plain. All "religious theories, schemes and systems which embrace notions of cosmogony . . . must . . . submit to the control of science, and relinquish all thought of controlling it. Acting other-wise proved disastrous in the past, and it is simply fatuous to-day."[38] The gauntlet had been thrown down. The city's atmosphere was electric and the local clerical fraternity hopping with rage as Tyndall's audience heard him preach a gospel of materialism.[39]

His address was a gunshot that echoed round the world. It prompted Bernard Shaw to put into the mouth of one of his characters, Mrs. Whitefield, the comment, "Nothing has been right since that speech that Professor Tyndall made at Belfast."[40] As Bernard Lightman has shown, it materially altered the public perception of Tyndall himself; before it, he had been portrayed in a positive light, afterward, as belligerent, conniving, and deceitful.[41] Events moved quickly. On Sunday, 23 August, Tyndall's address was the subject of a truculent attack by Rev. Professor Robert Watts at Fisherwick Church in downtown Belfast. Watts, the Assembly's College professor of systematic theology, had good reason for spitting blood. He had already submitted to the organizers of the Biology Section of the Belfast British Association meeting a paper congenially entitled "An Irenicum: Or, a Plea for Peace and Co-operation between Science and Theology."[42] They flatly rejected it.[43] It must have seemed to Watts that the scientific fraternity wasn't interested in peace and that hostilities had been initiated by "enlightened" scientists, not "entrenched" clergymen. For the spurned lecture—which Watts, not prepared to waste good words already committed to paper, delivered at noon the following Monday in Elmwood Presbyterian Church—revealed just how enthusiastic he could be about science.[44] But the British Association's rebuff *had* stung, and chagrin over his expulsion from the program had put him in a bad mood. Yet that was nothing to the anger that Tyndall's address aroused in him. The Sunday after the address, Watts turned his big guns on Tyndall in a sermon preached to an overflowing evening congregation at Fisherwick Place Church in the center of Belfast. The irenic tone of the rejected paper was gone. Tyndall's aim, according to Watts, was nothing less than the "extirpation of the Jehovah of the Bible." His mention of Epicurus was especially galling; that name had "become a synonym for sensualist," and Watts balked at the moral implications of adopting Epicurean values. It was a system that had "wrought the ruin of the communities and individuals who have acted out its principles in the past; and if the people of Belfast substitute it for the holy religion of the Son of God, and practise its degrading dogmas, the moral destiny of the metropolis of Ulster may easily be forecast." If Tyndall-type science was morally bankrupt, its scientific standing was no more secure. The "Darwinians," Watts reported, had nothing to say in response to the refuting evidence of reversion to type; all they could do was trade in the bogus currency of "unverified hypotheses." In sum, "Darwinism . . . notwithstanding its pretensions" was simply "unscientific" through and through.[45]

Details of Watts's attack were fully reported in the following week's edition of the *Witness*, and they proved so popular that the homily soon appeared as a pamphlet.[46] Within a month five thousand copies had been sold and a second edition called for.[47] The *Witness*'s insistence that the tract constituted "the most notable contribution to the controversial literature that has sprung out of the proceedings of the Belfast meeting of the British Association" enlarged at once Watts's reputation and the pamphlet's circulation.[48] Moreover, Tyndall and Huxley's refusal to engage Watts in a local public debate only added to *his* standing and to *their* notoriety in Calvinist Belfast.[49] "Perhaps at no time since the Arian Controversy," Rev. R. Jeffery reflected in October, had "the religious mind of Ulster been so deeply and indignantly stirred."[50]

Watts was not a lone voice imprecating Tyndall-style science. Rev. William Macloy put pen to paper to complain that Tyndall had evacuated the creation of its Creator and that "Darwin's boasted law of selection cannot furnish a single specimen of the evolution of a new species." "All species of plants and animals," he went on, "are explicable only by direct creative acts on the part of the great First Cause of all things."[51] As Watts was berating atomism before the downtown congregation of Fisherwick Place, the same message was buzzing through the ears of other congregations.[52] Rev. John MacNaughtan at Rosemary Street, Rev. George Shaw at Fitzroy, and Rev. T. Y. Killen at Duncairn all took up the cudgels.[53] In one way or another, Tyndall's intervention caught the attention of Anglicans and Unitarians, too. The Church of Ireland canon—and later bishop—Charles Parsons Reichel delivered his thoughts on the subject at both Carnmoney and St. Thomas's Parish churches on the subject; so, too, did James C. Street (who was decidedly more positive) and J. Scott Porter before Unitarian congregations.[54] Tyndall had certainly caught the mood. As he put it to a close friend, "Every pulpit in Belfast thundered of me."[55]

Overall, the tone of the sermons was dark. MacNaughtan, the Scottish minister of Rosemary Street, for example, closely followed Watts's diagnosis. He castigated the egregious address and inveighed "against [Tyndall's] erroneous teaching" as he sought to put his flock right on "Christianity and Science." Again the sermon was printed, this time at the request of his Kirk Session.[56] MacNaughtan's strategy to avert hostilities was to declare that "science and Christianity have different spheres" and that therefore "there is no reason why they should ever come into collision." But there was more. The speculative materialism that Tyndall had displayed *in excelsis* at the British Association was not merely bad science; it was contemptible moral philosophy. What hope, the preacher mused, could cold materialism give a human sufferer in the final throes of some fatal disease?[57]

These gut reactions to Tyndall's ambush disclose the tenor of Belfast's Calvinist reading of the Darwinian project in the aftermath of the British Association's visit to the city. The rhetorical space that his onslaught had opened up largely defined the cognitive zone in which the local debate about the new biology would be carried out for at least a couple of generations. What the community could *say* about Darwinism, and what they could *hear*, was profoundly shaped by the contingencies of the British Association event. Tyndall's speech had set the tenor and tempo of the Darwinian conversation in the city. And if the tone of the talk was aggressive or ill mannered, well then, it was Tyndall who was to blame. For observers didn't hesitate to point out that the association's president had violated all decorum.[58] As a local almanac for 1875 put it in its report of the meeting: "Professors Tyndall and Huxley . . . exhibited very bad taste and less sense in propounding absurd theories on the subject of creation . . . the views of Tyndall and Huxley are puerile and self-contradictory, opposed to common sense, and unsupported by any of the testimonies of history." Something of this adversarial atmosphere was captured in its representation of Rev. Professor Robert Watts:

> This year I faithfully define,
> A learned and orthodox divine
> Of wide and well deserved fame,
> Worthy the Presbyterian name.
> He can uphold our Banner Blue,
> And break a lance with Tyndall too,
> Exposing his fallacious rules
> Respecting atoms—molecules;
> O'erthrows all Huxley's speculations,
> And Darwin's vain imaginations;
> While Spencer's school received a share
> Of our Professor's watchful care.[59]

Rev. William MacIlwaine, whose stance on revivalism has already attracted our attention, was also irked by Tyndall's flouting of civic etiquette. In his own presidential address to the Belfast Naturalists' Field Club that winter, he complained that Tyndall's speech had been "reckless" and "a violation of the rules of good taste." He had abused privilege when he committed the "impropriety of debating a religious question in a scientific arena." And, as "one of the high priesthood of science," he had behaved shamefully by propagating materialism and atheism under the guise of science and delivering an oration at once "derogatory to Christianity" and "most unphilosophical."[60] The memory lingered long.

Nearly thirty years later, when the British Association returned to Belfast in 1902, Todd Martin recollected how the "placid waters were troubled for many days" when "Dr Tyndall . . . took advantage of his position as President to demand in the name of science as its inalienable right complete freedom in speculation and teaching, and . . . illustrated his claim by expounding a materialist theory of the universe."[61]

For all the consternation stirred up by Tyndall's transgressing science's frontiers into the heartlands of religion, there was a sense that this act of aggression was ultimately to be welcomed, for it displayed to the world the machinations of the materialist school. The gloves were off—and that was evidently a good thing! It was precisely because Tyndall had "spoken so plainly" that the editor of the *Witness* could happily observe, "We now know exactly the state of matters, and what is to be expected from Professor Tyndall and his school, and we shall be able to take our measures accordingly."[62]

The Winter of Discontent

When the Belfast Presbytery met for its monthly meeting on Wednesday, 2 September 1874, Rev. William Johnston, a recent moderator of the Church's General Assembly, rose to address the gathering. The principles of Christianity, he began, had just been subjected to an "open and determined onslaught." Two weeks ago that very day Professor John Tyndall had in the name of science assailed "Christianity in its vital parts." It pained him, Johnston went on, to hear the professor thanked for his disquisition and to discover, as had some of the newspapers, that "in Belfast such materialism and atheism [could be] propagated without any protest on the part of the audience who heard it."[63] But what bothered Johnston even more was that while the British Association had pronounced its blessing on Tyndall's speculations, that same body "had excluded a paper proposing to show how the inquiries of science could be conducted in thorough accordance with the principles of Christianity." "Thereby," he went on, "the Association made themselves a party to a one-sided attack on Christianity." Cries of "Hear, hear," from the assembled company indicated that he had touched a raw nerve among the Presbyterian ministers present. The paper to which Johnston referred, of course, was by Watts, and before concluding Johnston "proceeded to eulogise" Watts's riposte.[64]

Johnston had no intention of leaving the matter as a mere record of proceedings in the presbytery's minute book. A couple of weeks later he met with a number of fellow clergy to lay plans for a course of evening lectures on the relationship between Science and Christianity to be given at Rosemary Street Church during the winter months.[65] Arrangements were hastily finalized, and the *Witness* advertised their commencement in its columns on 27 November 1874. In time, these addresses would be drawn together into a book, distributed on both sides of the Atlantic, under the title *Science and Revelation: A Series of Lectures in Reply to the Theories of Tyndall, Huxley, Darwin, Spencer Etc* and furnished with a preface by Johnston himself. "Courtesy and precedent," Johnston noted in the preface, "forbade any protest at the time."[66] Now it was time for action.

During that 1874–1875 winter of discontent, eight Presbyterian theologians and one scientist—David Moore of the Glasnevin Botanical Gardens in Dublin[67]— joined together to stem, from the Rosemary Street pulpit, any materialist tide that Tyndall's rhetoric might trigger. Watts and MacNaughtan were, of course, among the coterie of apologists. But the first lecture, on "Science and Revelation: Their Distinctive Provinces," was delivered by J. L. Porter. In matters of science and religion, Porter was certain that good fences make good neighbors. Porter was perfectly happy to allow that "no theological dogma can annul a fact of science," but in an atmosphere redolent with the lingering aroma of Tyndall's presence, he was moved to castigate "crude theories and wild speculations," of which Darwinism was a prime case. As a piece of natural history, he conceded, *The Origin of Species* was "one of the most important contributions to modern science;" in logic, it was "an utter failure." Because its author persistently confused fact and theory, the book—however "striking" or "original"—had to be judged "not scientific." Species transmutation had never been directly witnessed, he insisted, and it was just simply reckless in the extreme to "set aside the Bible narrative" and assign "to man a common parentage with the monkey and the worm."[68] Darwinism, moreover, spooked Porter on another front—education. Pedagogy was important to him; in fact, it dominated his life. When he resigned his chair at Assembly's College in 1878, it was to take up the post of assistant commissioner of intermediate education, and, later, to become president of Queen's College.[69] So it was not surprising, as we have already seen, that in his opening address to the Presbyterian students for the session 1874–1875 he should pass judgment on the malevolent influence of recent scientific speculations. The spread of the Huxley-Tyndall cosmogony—though it was not real science—threatened to extinguish every virtue and trigger social breakdown. The need for theological colleges was thus more urgent than ever so that "heavenly light is preserved and cherished."[70]

The concern for wider social and moral matters was also uppermost in William Todd Martin's winter address a few days before Christmas. Earlier, in 1863, we recall, he had expressed grave concerns over the inroads being made by *Essays and Reviews* and Pentateuchal criticism. What troubled him now was the influence the idea of a Spencerian Impersonal Force would have on morality and religion—and thus on the foundations of the established social order. The new philosophy, he insisted, proposed "the founding anew of society" and the reconstruction of "the whole fabric of personal and social life." This reconstitution was the ultimate political goal of scientific naturalism, and it would be achieved by eradicating the idea of sin from human consciousness. As a consequence, morality would be reduced to a mere survival strategy. Vice was eliminated; the pleasure principle reigned supreme. And that was not all. A Darwinian society was not merely one in which moral conscience had been anesthetized; it was one in which all sorts of scary practices could be legitimated in the name of science. Already detecting the sorts of policy that later eugenicists would advocate, he urged: "A State free from the 'theological bias,' and in the hands of philosophic legislators, would offer a tempting field for experiment in the direction of a higher development of organism and intelligence, by careful scientific oversight of the question of population. Utilitarian ethics could facilitate this great enterprise by abolishing the Christian sentiment which protects the purity of the family."[71]

If Tyndall's most recent excursion into public controversy was the main stimulus for the winter lectures, his reduction of prayer to the language of energy was the subject of concern to Henry Wallace.[72] Back in June of 1872, Tyndall had received an anonymous letter from a member of the Athenaeum Club challenging religious believers to subject prayer to experimental analysis.[73] The Prayer Gauge Debate, as it came to be known, attracted numerous commentators, among them Francis Galton who achieved notoriety for his attempt, as Theodore Porter describes it, "to crush mystical piety under a heap of miscellaneous statistical facts."[74] Belfast Presbyterians were hardly prepared to stand idly by and witness prayer strangled by statistical methods and actuarial tables. Wallace, speaking in the Rosemary Street series on 18 January 1875, felt the need to search for some means of preventing the relentless laws of nature choking to death the age-old Christian practice. Prayer needed some maneuverability in a law-governed universe, and Wallace worked hard to give it the scientific and theological space it needed.[75]

The overall shape of the winter lecture series project is plain. Throughout the exercise, the intellectual elite of the Belfast Presbyterian community was determined to regain control of the debate by setting the terms in which the conversation about evolution had to be conducted. In the post-Tyndall era, evolution came to be seen as irretrievably wedded to materialism; as such it required censure on every front. So it was necessary to resurvey the major theological landmarks in Presbyterian territory and to reaffirm those boundaries. Just as the villagers of medieval Europe and colonial New England annually beat the bounds—marked out the village boundaries—so the Presbyterian hierarchy needed to reestablish its theological borders. Indeed, it is precisely for this reason, I suspect, that the Rosemary Street lectures included addresses—by A. C. Murphy on proofs of the Bible as divine revelation, by John MacNaughtan on the reality of sin, by John Moran on the life of Christ, and by William Magill on the divine origin of scripture—that made no specific reference to Tyndall, Darwin, Spencer, or any version of evolutionary theory. That they were included in the series attests to the perceived need to systematically reaffirm the cardinal doctrines of the faith so as to ensure that Presbyterian theological territory remained intact.

Bishops' Move

If Belfast Protestants found the British Association's recent meeting repugnant, Catholics had no less cause to squirm. Within days of the Belfast meeting, Huxley would be describing the anonymous attack on *The Descent of Man*—by St. George Mivart, the Catholic evolutionist—as replete with the "misrepresentation and falsification [that] are the favourite weapons of Jesuitical Rome" and thus all too typical of "the secret poisonings of the Papal Borgias." With Darwin's complicity he schemed . . . and secured "Mivart's excommunication from the church scientific."[76]

Just over a year earlier, the Catholic serial *The Irish Ecclesiastical Record* had presented an evaluation of "Darwinism" in which its correspondent castigated the "Moloch of natural selection" for its "ruthless extermination of . . . unsuccessful competitors," for its lack of evidence for transitional forms, for its failure to account for gaps in the fossil record, and for its distasteful moral implications.[77] As for the latest clash in Belfast, the Catholic archbishops and bishops of Ireland issued a pastoral letter in November 1874 in which they repudiated the "blasphemy upon this Catholic nation" recently uttered by the "professors of Materialism . . . under the name of Science." There was nothing new about this tactic, the Catholic hierarchy reflected, but they did think that this most recent incarnation of materialism unveiled more clearly than ever before "the moral and social doctrines that lurked in the gloomy recesses of [science's] speculative theories." Quite simply, it meant that moral responsibility had been erased, that virtue and vice had become but "expressions of the same mechanical force," and that sin and holiness likewise vanished chimera-like into oblivion. Everything in human life from "sensual love" to religious sentiment were "all equally results of the play between organism and environment through countless ages of the past." Such was the brutalizing machine that now confronted the Irish people.[78]

The congruence between these evaluations and those of the Presbyterian commentators we have scrutinized is certainly marked. Why then should Watts, in a subsequent reprint of his pamphlet "Atomism: Dr. Tyndall's Atomic Theory of the Universe Examined and Refuted" incorporate in an appendix "strictures on the recent Manifesto of the Roman Catholic Hierarchy of Ireland in reference to the sphere of Science"? That he wished to distance himself from their proposals is clear. It was "painful," he noted, "to observe the position taken by the Roman Catholic hierarchy of Ireland in their answer to Professors Tyndall and Huxley." So, just what was the problem? Simply this—Watts felt that the Catholic willingness to compartmentalize science and religion would lead to "the secularisation of the physical sciences." Because Cardinal Cullen and his coreligionists seemed only too willing to follow Newman in restricting the physicist to the bare empirical, Watts insisted that since "the Word of God enjoins it upon men as a duty to infer the invisible things of the Creator from the things that are made," the Roman Catholic hierarchy of Ireland had, by their declaration, "taken up an attitude of antagonism to that Word, prohibiting scientists, as such, from rising above the law to the infinitely wise, Almighty Lawgiver."[79]

Two of Watts's close Presbyterian allies in the anti-Tyndall campaign, however, had taken this same line. J. L. Porter, in the first lecture of the winter series, had appealed for a clear-cut boundary line—in terms of content and methodology—between the provinces of science and theology. For him, it was a case of reconciliation by segregation; good relations required careful boundary maintenance.[80] Again, in an attempt to avert hostilities, Rev. John MacNaughtan, in his post–British Association sermon, insisted that "science and Christianity have different spheres" and that "there is no reason why they should ever come into collision." The appropriate allocation of items to the respective regional geographies of science and religion would greatly assist in keeping the peace.[81]

The sectarian traditions in Irish religion doubtless had a key role to play in these particular machinations. The commotion surrounding the Tyndall event merely became yet another occasion for Ulster nonconformity to uncover its sense of being under siege.[82] Watts had no desire to share his doctrinal space with Catholic bishops who were part of a tradition that had all too recently poured scorn on the New York meeting of the Evangelical Alliance and attacked his own hero, the Princeton theologian Charles Hodge. Moreover, though he may have privately agreed with their dismissal of the Alliance "debate on the Darwinian theory"—in which McCosh had participated—as "empty—nay, . . . almost childish," it was clearly important to Watts not to be seen in the company of such critics.[83] In his confrontation with evolutionary theory, Watts wanted to cultivate and tend to his own tradition's theological space and not engage in extramural affiliations. And by seeking to cast secularization and Catholicism as coconspirators against the inductive truths of science and the revealed truths of scripture, he found it possible to conflate as a single object of opprobrium the old enemy—popery—and the new enemy—evolution.

For their part, the Catholic hierarchy did not miss the opportunity of firing its own broadsides at Protestantism. The bishops and archbishops thought it would "not be amiss, in connection with the Irish National system of education, to call attention to the fact, that the Materialists of to-day are able to boast that the doctrines which have brought most odium upon their school have been openly taught by a high Protestant dignitary." It only confirmed them in their uncompromising stance on the Catholic educational system. Had it not been for their vigilance, an unbelieving tide would have swept through the entire curriculum. Such circumstances justified "to the full the determination of Catholic Ireland not to allow her young men to frequent Universities and Colleges where Science is made the vehicle of Materialism." Accordingly, the bishops and archbishops rebuked "the indifference of those who may be tempted to grow slack in the struggle for a Catholic system of education."[84] Tyndall's speech, it seems, succeeded not only in fostering the opposition of Protestants and Catholics to his own science but also in furthering their antagonism to each other.

No Surrender

In the wake of Tyndall's offensive, the Calvinist community in Belfast found it virtually impossible for at least a generation to find any rapprochement with Darwinian biology. One moment, a decade after the notorious British Association meeting, serves to illustrate how persistently antagonistic sentiments lingered on in the city—compared with voices from other locations. The pan-Presbyterian General Council met in Belfast during the last week of June 1884. At this gathering, several speakers called for the incorporation of evolution theory into the fabric of Calvinist theology. The Scottish clergyman, hymnologist, and author George Matheson, for example, told his hearers that he could detect the hidden hand of divinity behind Spencer's "Unknowable," a viewpoint that, as we have seen, he further elaborated on the following year in *Can the Old Faith Live with the New? Or, The Problem of Evolution and Revelation*. Spencer's Force, as Matheson encountered it, was none other than the Spirit of God and the doctrine of Evolution nothing less than a revelation of the workings of Providence. Indeed, he was prepared to suggest that "in the production of the human race there may have co-operated the factors called Natural Selection, Heredity, Concomitant Variation, and Environment."[85] Professor Jean Monod of Montauban, France, who followed Matheson on the program, concurred in his judgments about the "possibility of harmonizing with the leading doctrines of Christianity, the system of Evolution rightly understood." The Scottish moral philosopher Henry Calderwood (supported by Professor Stewart D. F. Salmond from Aberdeen) threw his weight behind Matheson's proposals, insisting that "Evolution theory is a theory teaching Creation."[86]

Local delegates, however, were not so sanguine. The implacable Watts told them outright that evolution was a "mere hypothesis" with "not one single particle of evidence" to support it. The need for embarking on any doctrinal reconstruction did not exactly commend itself to him.[87] Indeed, he had long been campaigning—in Belfast and in Scotland—against Spencer's system because to him it was explicitly designed "to overthrow the Scripture doctrine of special creations."[88] To hear latter-day Scottish Calvinists divinizing the "Inscrutable Ultimate" must have sounded like the grossest act of doctrinal betrayal. To reconstruct theology along Matheson's lines would obviously be flagrantly irresponsible.

Others concurred. A year or two later, in the shadow of Matheson's flirtation with a Spencerian-honed theology, Todd Martin issued a book-length criticism of Spencer's cosmic system expressing typically inductivist suspicions of the entire project. Cosmic evolutionists, he judged, were forever going on about their far-flung empire and the comprehensiveness of their claims, and thereby they persistently committed the cardinal error of elevating the particular into the universal. Their claims simply had to be relegated to the class of spurious, untestable hypotheses. If ever there was a group who routinely read their own metaphysics into the data, it was the cosmic evolutionists. It was a serious state of affairs. "The highest interests are at stake," he declared. "Evolutionism, if accepted, must eventually crush the liberty of the spirit in man; and the liberty of the spirit is indispensable to the progress of humanity." In particular, Martin was concerned about the destructive inroads evolutionism would have on human morality and epistemology. Here he advanced a novel argument about evolution's incapacity to deliver a persuasive account of human cognizing.[89] On the evolutionary hypothesis, Martin noted, the sense organs had developed for the purpose of adapting the human species to its environment. But the human capacity to deliver true knowledge of an advanced kind (including the theory of cosmic evolution) was a different matter altogether, going well beyond any mere survival function.

> It is not clear, then, how sense organs evolved for a quite different end can be relied on to give such a full and complete knowledge of the phenomena as will furnish a basis for a perfect cosmic theory . . . The senses with which we are endowed are, on

the hypothesis of evolution, defective in their adaptation to any purpose except the practical end of adjustment to environment, and so to the maintenance of man's life on the earth. Admirably fitted to serve the purpose for which they have been evolved, they fail us when applied to any other use. They cannot, therefore, be depended on as instruments of exact knowledge.[90]

Morally perilous and epistemically bankrupt as it was, he could only respond to his self-directed question, "Is Evolutionism true?" in one way: "The answer is a decided negative."

In the aftermath of the pan-Presbyterian gathering in Belfast, the ever-watchful Robert Watts continued to keep an eagle eye on things. He increasingly retreated to an ever-more polarized position. Comparing what he called "the Huxleyan Kosmogony" with the Mosaic record, he insisted that, "According to Moses, the several species were brought into existence by distinct creative acts, not *mediately* through the utilisation of previously existing organisms modified and adapted to new ends and habitats." Indeed, it was clear to him that the whole tenor of the Mosaic account was in "antagonism to our modern evolutionists."[91] The Genesis narrative was solid ground, not to be compared with those "imaginary intermediate organic forms" conjured into existence by "Dame Fancy."[92]

By now, Henry Drummond's *Natural Law in the Spiritual World* had already fallen under Watts's gaze, and as soon as *The Ascent of Man* appeared on the horizon Watts had it in his sights.[93] Drummond's extravagant evangelical efforts to evolutionize theology and to import Spencer's evolutionary laws into the supernatural realm created a sensation at the time, and overall reactions were remarkably sanguine.[94] But, as we have already seen, there were critics. Horatius Bonar, the Earl of Shaftesbury, and the conservative Highland flank of the Free Church of Scotland all abominated Drummond's project. Watts concurred: the Drummond project was just a mess—theologically and scientifically. To be sure, Drummond's aim of removing the "alleged antagonism" between science and religion was a worthy objective. But his strategy in the last analysis was "not an Irenicum between science and religion or between the laws of the empires of matter and of spirit" at all. Rather, it only displayed the expansionist character of natural law. Watts found the whole project so bizarre that, as he worked up to his conclusion, he felt "inclined to apologize for attempting a formal refutation of a theory which, if it means anything intelligible, involves the denial of all that the Scriptures teach, and all that Christian experience reveals."[95] Drummond's evolutionism disturbed him in other ways, too, not least on matters to do with gender relations. The priority that Drummond had allocated to the evolutionary role of motherhood in social evolution upset the settled cosmos of Watts's brand of Calvinism. To him, Drummond's "glorification of motherhood" was "neither scriptural nor scientific" and undermined the primacy entrusted to fathers by God and nature. Scripture, he insisted, gave "fatherhood the precedence as the fontal source, without which there never could have been either motherhood or childhood." Drummond's evolution disturbed the foundations of a domestic social order in which "fatherhood must ever hold a place as much superior to motherhood, as the life-germ holds in relation to its habitat."[96]

Watts's disgust at Drummond's evolutionary metaphysics was matched only by his revulsion at the biblical criticism of Robertson Smith. In refusing to countenance higher criticism in any shape or form, Watts was giving voice to the long-standing opposition of conservative Ulster Calvinists to the entire enterprise. Since the 1830s, a succession of local professors had mounted resistance to what they saw as German infidelity. In their defense of plenary inspiration, they mobilized Common Sense suspicions about speculative theory over against the plain meaning of texts to shore up traditional confessionalism.[97] In the early 1860s, J. G. Murphy voiced his opposition to any suggestion that contradictory sources from different documents had been put together under the editorial control of a redactor. His readers learned that there was no need to project such inventions; "all the diversities of style that have been or can be discovered . . . do not suffice to prove a work to be a medley from different authors."[98] Watts, who stood squarely in that old school tradition, was profoundly disturbed by any effort to seek rapprochement with the newest trends and remained doggedly determined to ferret out concessions wherever he could find them.

Not surprisingly, Robertson Smith alarmed him to the core, and in 1881 he took over three hundred pages to voice his criticisms. The critique was popular among certain audiences, and the book rapidly went into two more editions. In preliminary observations marked by their fair share of snide commentary, Watts reduced the entire business to a single issue; in the work of Robertson Smith, he remarked, "We are brought face to face with the fearful alternative of accepting as the word of God a palpable forgery claiming to be divinely inspired, or of rejecting it as a mockery and a fraud." In short, the whole operation was "faith-subverting." As for those who claimed that Smith was the one theologian whose scholarship was sufficiently robust to counter the radical enemies of the faith, Watts quipped that the most appropriate "precedent for such 'speaking with the enemy' would be the conference held by Lundy, governor of Derry, with the emissaries of King James, when he agreed to surrender to that monarch the Maiden City."[99] Robert Lundy, the governor of Derry, it should be noted, is still despised among Ulster

Unionists as a traitor and is annually burnt in effigy during the celebrations marking the siege of Derry.

Not surprisingly, Watts had grown entirely disillusioned with the compromises of the Scottish Presbyterians, not least the Edinburgh New College network, and he continued to cultivate links instead with Princeton Theological Seminary in New Jersey. In introducing a student to the Princeton campus, he wrote to the theologian B. B. Warfield in 1893: "I am greatly pleased to find that our young men have turned their eyes to Princeton instead of Edinburgh."[100] Later that same year, he concluded yet another letter with the comment: "I dread the influence of the Scotch Theological Halls."[101] He had in mind figures like Marcus Dods, A. B. Bruce, and, of course, Robertson Smith, whose critical scholarship he found outrageous.[102] For his part, Dods was appalled and reportedly quipped that Watts was "one of those unhappily constituted men who cannot write unless they are angry."[103] Watts would not have been surprised at such sniping, yet he can't have been overjoyed to read the review of his book on Robertson Smith that William Henry Green, the distinguished Princeton professor of biblical and Oriental literature, produced in 1882. While Green issued some soporific comments on the theological appropriateness of Watts's attack, he plainly considered that Watts's critical scholarship fell far short of the mark. The book, Green remarked, "does not furnish a complete answer to the hypothesis in question, and does not even undertake to grapple with the critical grounds adduced on its behalf." It was marred by a "precarious" line of reasoning here, a "not very obvious" charge there, and arguments that could "scarcely be considered conclusive."[104] Theological predilection had evidently triumphed too fully over serious scholarly rebuttal.

As Watts kept up his assault on the newest impulses in scientific and biblical scholarship, the memory of the visit of the British Association to his home city continued to burn bright. A full twenty years after Tyndall's infamous Belfast Address, Watts was still recalling its repercussions to his Princeton correspondent: "Immediately after the meeting of the British Association in Belfast, in 1874, I reviewed Spencer's Biological Hypothesis, & proved that he was neither a philosopher, nor a scientist . . . Dr. McCosh's successor in the Queen's College here, trots out Spencer to our young men in their undergraduate course, and one of my duties . . . is to pump out of them Mill & Spencer."[105] Plainly, Watts neither would, nor could, release his grip on the bitter memory of that episode.

It lingered too in the memory of the medical practitioner Samuel B. G. McKinney of Sentry Hill—the family home of eminent Presbyterian gentleman-farming stock located just to the north of the city. In an 1895 work that was printed in several editions, McKinney reprised many of the objections to evolutionism that had wended their way through the winter series of lectures more than twenty years earlier.[106] McKinney, who had been a student at the Queen's Colleges and then in Edinburgh during the late 1860s and early 1870s and later the author of several theological works, clearly rejected evolution: "The weakness of the theory lies in the absence of any foundation of fact."[107] He went on: "There is not a particle of scientific evidence that a particular kind of protoplasm which develops into an elephant can ever . . . lose its power to produce an elephant and gain the power of producing a different animal."[108] Several years earlier in 1888, he had resorted to phrenology to launch a rather offensive attack on Darwin himself.

A glance at any photograph of Darwin is sufficient to convince any one that his brain was so imperfectly developed that he was not naturally capable of exhibiting any higher functions of mind, and could only be a keen observer of facts and a steady plodder in experiments. Even his experiments on the influence of worms were due to the suggestions of another, and he originated nothing . . . Although the evolution theory was contrary to reason and to scientific principles, the imperfection of his brain and the deficiency of his education in the knowledge of perfect archetypes made Darwin incapable of feeling the full force of the absurdity of his notion that poverty and ignorance are scientific indications of natural inferiority.[109]

Of course, concerns about teleology, morality, and the links between mind and matter also typically intruded. But McKinney went further, drawing on his travel experience to accuse evolutionary anthropologists of racism and imperialism. "The theory of evolution is comforting to the slave-owner and the despot," he announced, "who argue that the control of inferior races is the birthright of the more highly-evolved." Indeed, his time abroad as a ship's physician during a number of voyages during 1877–1879, which included a winter visit to the Cape of Good Hope, had sensitized him to the qualities of races too comfortably relegated to the margins of evolutionary progress by Darwinian enthusiasts.

Englishmen used to fancy that an inferior race dwelt in New Zealand; but a few wars with the Maories [sic] proved that it is safer to boast of superiority in the security of an English study than on the field of battle. The scientific world was convinced that the New Zealander could not be many millions of years behind in evolution. There remained the consolation that the Negro may be bullied and insulted with impunity; and failure to resist tyranny is the great evidence of lowness of type according to modern teaching . . .

Ignorant residents of European cities speak with great contempt of Hottentots as filthy wretches who hardly ever wash; but those accustomed from infancy to be waited on by servants, and to use sewers which they never see, and to find a bath easily obtained by turning a tap, are little better than children in a nursery as compared with independent men.[110]

Although written while in medical practice in London, where he moved in 1882, McKinney's reading of evolutionary theory was a compound product of his Belfast Presbyterian heritage and his experience of overseas travel; both coalesced to reinforce a staunchly anti-Darwinian stance.

Counterpoint: Derry and Dublin

The rhetorical fervor that typified the Belfast contact zone sharply contrasts with the tone adopted by John Robinson Leebody, professor of mathematics and natural philosophy since 1865 at the Presbyterian Magee College outside Londonderry. During the 1870s, Leebody issued several commentaries on the new visions of nature and religion emanating from the pens of Darwin, Ernst Haeckel, Henry Maudsley, Lionel Beale, John William Draper, Arthur Balfour, and St. George Mivart. Before the Belfast rumpus, he had reviewed Tyndall's 1870 discourse to the Liverpool meeting of the British Association in a talk entitled "The Scientific Uses of the Imagination." Taking the opportunity to rehearse the views of Pierre-Simon Laplace, Auguste Comte, Edward Burnett Tylor, Lubbock, Spencer, and Karl Vogt, as well as Tyndall, Darwin and Ernst Haeckel, Leebody carefully discriminated various modes in which the theory of evolution might be held. He objected to its most materialistic rendition, which asserted a radical continuity between matter and mind and rejected attempts to reduce "ideas of virtue, truth, and God" to "fictions of the mind, evolved by the ceaseless activity of human thought." Less imperial versions, however, were a different matter:

> Stated with these restrictions, we do not see that the doctrine of evolution comes in contact with the teachings of Scripture at all, however it may conflict with traditional preconceptions which have become bound up with our religious beliefs. . . . "It is," says Dr McCosh, speaking of its application to account for the origin of species amongst lower animals, "a question to be decided by naturalists and not by theologians, who . . . have no authority from the Word of God to say that every species of tiny moth has been created independently of all species of moths which have gone before."

Thus, even while objecting to Darwinian morality, Leebody welcomed "in the interests of both Science and Religion, . . . the appearance of the 'Descent of Man.' It enriches science by a vast number of valuable facts, and it will stimulate inquiry with regard to the theory of Evolution which may be expected to yield important results."[111]

With these perspectives already in place *prior* to the theatrics of the Belfast meeting of the British Association, Leebody seems to have been much less shaken by the Tyndall onslaught. Writing in 1876, in a critical commentary on efforts to prosecute what he called "the principle of continuity" and the "correlation of forces," he paused to issue a far more positive commentary on the Darwinian vision than anything expressed just two years or so earlier in the 1874 winter series of lectures in Belfast: "The more fully . . . the animal and vegetable worlds are examined, the more fully is the value of Mr Darwin's investigations and speculations seen. . . . [T]here is no one at all familiar with the history of science and the present tendencies of scientific discovery who does not see that Mr Darwin's name is one destined to stand in the first rank among the leaders of intellectual progress. The Newton of biological science has yet, we believe, to arise, but scientific men are pretty generally agreed that Mr Darwin must at least be regarded as the Kepler."[112]

No doubt there are parallels to be detected between Leebody's anxieties over materialist reductionism and those of his Belfast colleagues. But his discrimination between scientific findings and philosophical speculation was much more carefully prosecuted; his language was much more restrained; his assessments of Darwinism were progressively more tolerant; his enthusiasm for science more thoroughgoing.[113] And he later came to find value in the application of evolution to the history of religion itself, conceding that "Professor Huxley does show that, to a large extent, various forms of mythology and false religion must be regarded as a natural outcome of the intellectual development of the race."[114]

Leebody plainly occupied a different rhetorical space from his Belfast colleagues. Several things seem to have contributed. As a trained scientist, he felt the need to keep scientific inquiry free from unwarranted theological policing. He was well aware, for example, "that some of our ideas with regard to creation and the past history of our planet have recently undergone a change, and that we cannot claim to have been infallible in our interpretations of the opening chapters of Genesis."[115] His concern to maintain dialogue between science and religion frequently obtruded:

> The history of the past tells us that we need not dread that the religious beliefs of
> the community will be enfeebled or destroyed by the advance of science. It may tend
> to displace some of the traditional beliefs which are the excrescences on Scripture
> truth rightly formulated, but that will be no loss. A century or two ago it destroyed

the belief in witchcraft, which up till that time was considered a crucial test in discriminating between a thoroughly orthodox theologian and one with dangerously rationalistic tendencies. In recent times it has taught us to be slow in interpreting literally portions of the Old Testament which unquestionably were not left on record as an exposition of cosmological science.[116]

No less significant was the educational space that Leebody occupied. The Magee College, which combined "secular" and "religious" education, enjoyed no state support. Leebody had no desire to change this circumstance but believed passionately that his students should be placed on an equal footing with other universities in Ireland by being admitted to the University of Ireland examinations for the conferment of a degree. In such a context, the need to promote nonsectarian scientific training was paramount, and Leebody resolutely distanced himself from a Catholic ideology that insisted that *all* subjects must be taught, as Cardinal Cullen put it, "on purely Catholic principles." Leebody vigorously protested: "There is no Protestant Mathematics or Chemistry as distinguished from that taught in a Catholic college."[117] In the educational culture wars of late nineteenth-century Ireland, Leebody had to occupy a rhetorical zone that was seen to preserve the independence of scientific inquiry from too much theological supervision.[118] Taken together, Leebody's location constituted a social space different from that produced among Belfast Presbyterians by Tyndall's pugnacious address.

A brief snapshot of another intervention—this time from Dublin—further serves to sharpen the contrast with circumstances in Belfast. Alexander Macalister (1844–1919), the distinguished Trinity College professor of anatomy (who from 1883 occupied the chair of anatomy at Cambridge for thirty-six years), Fellow of the Royal Society, and prominent Presbyterian layman, was a key member of a scientific circle rotating around Trinity College and the Royal College of Science. That circle included such luminaries as Ireland's astronomer royal Robert Ball; the botanist W. T. Thiselton-Dyer, who taught there prior to moving to Kew Gardens; and the anthropologist Alfred Cort Haddon, who took up the chair of zoology at the College of Science in the early 1880s and during his tenure led the celebrated Torres Expedition of 1898–1899. All were advocates of natural selection—a marked departure from the earlier opposition of Samuel Haughton, the Trinity professor of geology, and the botanist William Henry Harvey.[119]

Given his professional expertise, it is not surprising that Macalister provided a Darwinian reading of anatomical variations, particularly vestigial organs, and sought evolutionary continuities between human and primate physiology. He had long been convinced that questions about evolution "must be decided by work in the dissecting-room, the field, the zoological garden, and the laboratory" and that the "shred of old prejudice against researches in the principles of biology" that continued to be entertained by some religious partisans was the preserve of the "narrow-minded." The failure of "teleological theory" and talk of archetypal plans to provide a persuasive account of certain aspects of animal morphology had led him in 1870 to announce that "at the present day there are few comparative anatomists who have not given their allegiance to evolutionism."[120]

Whatever he thought about the scientific inadequacies of teleology, as a prominent Presbyterian layman, Macalister affirmed his faith in design in a forty-eight-page pamphlet published by the Religious Tract Society in 1886 in which he used the human body as the vehicle for confirming divine purpose. After a detailed excursus on the mechanical glories of the human skeleton, heart, nervous system, muscles—in which the doxological rhetoric of the wonderful and marvelous prevailed—he concluded, "The prominent lesson which the examination of the human body impresses on us is that of perfect adaptation of means to ends, of structure to function. The unprejudiced mind cannot fail to read in every organ . . . the inscriptions of purpose, and to learn thereby that they are the products of supreme power directed by supreme wisdom."[121] If teleology had no place in scientific explanation, it still occupied a prominent location in human meditation on the natural order. Indeed, this tract for the times was, in many ways, an expansion of a sermon he had delivered at the annual meeting of the Jervis Street Mission Church fifteen years earlier in 1871.[122]

Macalister's own accommodation to Darwinism had long been worked out. In an extended review of Darwin's *The Descent of Man* in 1871, for example, he concluded that, as far as the human corporeal frame was concerned, "the defenders of evolution have the best of evidence on their side." For him, critical confirmation came from human rudimentary organs; as he put it, "how to account for these rudiments on any other but an evolution theory it is very hard to see. No teleological reason for their existence can be given." Darwinian accounts of the human psyche, however, were a different matter. Here Macalister insisted on a fundamental dualism in human psychical nature—what he called the "duality of the Ego": "one [part], the seat of the passions, desires, and appetites is identical with that of the lower animals, and in this part subsists all the feelings which Mr. Darwin relies on to prove the derivative nature of man's rationality; the other is the part which has no correlate in the lower animals, the seat of the moral sense, and the religious feelings . . . which no evolution can account for."[123] Macalister referred to this accommodationist strategy as "mixed evolution," namely, allowing evolution as a persuasive explanation for certain aspects of the human species while exempting other dimensions, notably moral sensibilities, from its power.

These latter reservations later drew dispiriting remarks from his obituarist, Grafton Elliot Smith (like Macalister an anatomist, anthropologist, and Egyptologist), about intellectual compromise and an absence of the "excitement" that energized contemporaries caught up in the full flush of evolutionary imperatives. In the pages of the Royal Society's *Proceedings,* Smith thus remarked that Macalister's addiction to the mere accumulation of anatomical facts "shed some light upon the remarkable fact that British anatomy entered upon a period of profound stagnation."[124] Smith was deeply committed to evolution as an all-encompassing ideology and, in 1913 was involved in the Piltdown controversy, pronouncing that the remains—later shown to be a hoax—constituted the most primitive human and simian brain ever recorded.[125]

As a leading specialist on the evolution of the primate brain—he was awarded the Royal Medal for his work on comparative neuroanatomy—it is hardly surprising that Smith found Macalister's evolution relatively timid. But Macalister's time in Ireland had exposed him to intellectual influences less fervent about progress. There, archaeological interest in the pre-Christian Irish past, dating back to the early nineteenth-century researches of George Petrie and Samuel Ferguson, had cultivated a nostalgic romanticism for earlier times and cemented the conviction that, whatever their material conditions, the early Irish were far from morally, culturally, or spiritually barbarous. Such virtues profoundly challenged radical evolutionary progressivism but did not undermine Macalister's ideas about evolutionary development. Still, like his Dublin and Cambridge colleague Haddon, who also enthusiastically pushed forward the work of the Dublin Anthropometric Laboratory, Macalister's progressivism remained comparatively muted.[126] Certainly, as he put it in his 1892 presidential address to Section H of the British Association for the Advancement of Science, he came to think that the study of craniometry had, for all its undoubted industry, delivered no coherent or satisfactory "unifying hypothesis."[127] Yet his reservations were not intended to deny the significance of Darwinian evolution. To be sure, Macalister's endorsement fell short of the evolutionary expansionism championed by Smith, but it was sufficiently robust to make the Trinity Anatomy Department "a redoubt of evolutionary biology" by the 1880s, despite the alarm that Tyndall's Belfast address had awakened among the Catholic bishops.[128] Cardinal Cullen, for example, is reported to have asserted in 1869: "If a man teaching Chemistry or Geology were to assert that the cosmogony of Moses was in opposition to the order of things at present existing I would remove him from his teaching." And it seems entirely significant that the Royal Commission on higher education put the following question to representatives of the Catholic medical school: "Would you give a Professor of Comparative Anatomy . . . a free hand in regard to Darwin's theory of Natural Selection?"[129]

Thus, while the *Tablet* railed against scientific rationalists in the universities and a growing Irish nationalism gave British scientists cause for concern over the future of the curriculum in the Irish higher education system, Macalister's own denomination allowed him space in 1882 to address the Dublin Presbyterian Association (of which he had earlier been president) on the subject of evolution in church history.[130] Seeking to encompass the development of ecclesiastical polity and dogmatic forms (what he also called "evolution of the Church organism") within the "operation of those great laws which work in other departments of the material and moral world," he used evolutionary vocabulary to speak of the ways in which ritual and organizational "variations" might be explained.

> When, in the history of any set of phenomena, we find that at certain successive stages the sequences vary, and that these variations are directly and recognizably related to the external surrounding conditions, we assume that there is a direct connexion between these external conditions and the modified sequences, and we say that the resulting state of things is due to evolution. The fundamental postulates are a capacity of variation in the train of sequences, and external modifying influences, and the latter may be either the direct action of the environments on the phenomena, or may be due to a power from without, overruling and directly ordering the modifications. In this sense evolution may be defined as the principle in accordance with which phenomena are modified to keep them in harmony with their surrounding conditions.[131]

Whereas in the North, Robert Watts poured scorn on Henry Drummond, Alexander Macalister warmly endorsed his project. His books, Macalister told the readers of the *Bookman* in 1897 at the time of Drummond's death, "attracted the public attention by their unique blending of the most thorough-going evolutionism with as thorough-going an evangelicalism, as well as by their fascinating literary style and their happy illustrations of the themes on which he wrote."[132] It was the same with his endorsement of James Iverach's evolutionism. Commenting on Iverach's volume of lectures, *Theism in the Light of Science and Philosophy,* he observed, "The main argument is well expressed and cogent, that evolution has proceeded along orderly lines, and there is nothing in the facts to exclude, and much that directly involves, predetermination and supervision."[133]

◆ ◆ ◆

TYNDALL'S BRITISH ASSOCIATION speech cast a long shadow over intellectual life in Victorian Belfast. The flavor of his "materialist manifesto" lingered on and on and made it far more difficult for local churchmen to negotiate the sort of conceptual arrangements being effected in other places. In such an atmosphere, they could only see the folly of acceding to the accommodationist strategies of a Drummond or a Matheson. These newer compromises only seemed to confirm them in the rightness of the belligerent path on which they had embarked a decade and more earlier. In other Irish settings—notably, Derry and Dublin—Presbyterians like Leebody and Macalister found it possible to issue far more positive commentaries on Darwinian evolution not only because of their scientific expertise but also because of their desire to maintain a curriculum unfettered by what they considered to be the excessive theological policing that characterized Catholic educational enterprises. The Belfast reading of Darwin, when set against the horizon of these alternative judgments, was thus a compound product of local circumstances: Tyndall's theatrics at the meeting of the British Association, tussles over who should control higher education in Ireland, and a long-standing anti-Catholicism that colored virtually every aspect of cultural and political life during the final decades of the nineteenth century.

CHAPTER FOUR

Toronto, Knox, and Bacon's Bequest

On Monday, 9 June 1884, the pages of the *Toronto World* were host to a spat between a certain Dr. Wild and an anonymous correspondent writing under the signature "Evolutionist." Dr. Joseph Wild, a theological controversialist and, since 1880, firebrand minister of the Bond Street Congregational Church in Toronto, was systematically stirring up his congregation on a number of his pet peeves in the late spring of 1884. Born in England in 1834, a British Israelite and apologist for the glories of the British Empire, he routinely attracted crowds of more than three thousand to his Sunday evening performances.[1] On 8 June, evolution was on his mind, and readers of the *Toronto World* were promised that "the Pope and Freemasonry" would be in his crosshairs the following Sunday. The journalist recorded that Wild "concluded by ridiculing the theories of some writers as to an evolution from an incandescent nebulae as being as silly as the evolution from a monkey."[2] A few days later, "Evolutionist" replied. Terse and to the point, he queried the reverend doctor's competence as a historian and a naturalist and recommended that he had better "confine himself strictly to theology." As for the future, the unnamed Darwinian went on, "The world may come to an end in 1935, but men of science cannot afford to renounce their belief in evolution."[3]

From time to time, since the 1870s, comparable reports of anti-Darwinian sentiments had found their way into the Canadian press. A decade or so earlier, when Ottawa's new Dominion Methodist Church opened its basement on 26 December 1875, a congregation of over a thousand turned up, despite a "fierce storm of sleet," to hear an address by the church's minister Rev. William J. Hunter. His subject for this particular evening was the contemporary arts, especially as they bore on the subject of what he called "The Giant Depravity of Man." In a short space of time, he was tackling the topic "Religion vs Science." Protesting that he had "no quarrel with true science, philosophy and literature," he hit out at "Darwin and his followers," who, he told his hearers, "claim monkeys as their ancestry," and at John Tyndall, whose profane querying of the efficacy of prayer was well known. "Let Darwin, and Tyndall, and Huxley, and the whole brood of modern free-thinkers prosecute their researches to the uttermost," he went on, "the Ark of God is not in danger."[4]

Dispatches also came from further afield. The November 1884 readers of the *Toronto World*, for instance, were treated to a piece entitled "Dr Talmage's Laughable Description of Evolution." It was from a New York correspondent reporting on an address by the sensationalist Brooklyn preacher and reformer Thomas De Witt Talmage, who had apparently declared, "Evolution is out and out infidelity. Paine, Hume and Voltaire no more disbelieve the holy scriptures than those who believe in evolution." To Talmage, the "dogma" of the "survival of the fittest" was particularly objectionable and showed just how "brutalizing" Darwin's theory really was. With its conception that "a man is a bankrupt monkey," Talmage concluded that evolution was nothing more than "an old heathen corpse set up in a morgue, and Darwin and Spencer have been trying to galvanize it."[5]

Closer to home, when Alfred Russel Wallace spoke at the University of Toronto during his North American tour of 1886–1887, the newspapers carried lengthy letters of protest from those who objected to him addressing the university on the subject of "Darwinism." Others wrote directly to Daniel Wilson, first president of the federated University of Toronto. One from a graduate of the university grumbled at the use of the University's Convocation Hall to promote "a theory having for its aim the overthrow of the doctrine of a Creator and the establishment of atheism." The complainant went on to elaborate his own understanding of the "mysteries of Creation" and unburdened himself of a set of speculations about the age of the earth that prompted the president to send him "a soothing little note" in hopes that he would "hear no more about it."[6] For his own part, Wilson had anticipated such problems and jotted down in his journal for March 1887 that Wallace's coming to the university was "driving the Baptists into setting up a University of their own . . . The very name of Darwin is to most of them like a red rag to a bull; and the greater their ignorance the more pronounced their dogmatism."[7]

It is tempting to assume that the casual remarks of two or three truculent clergy, or the feather ruffling occasioned by Wallace's presence at Toronto's center of learning, are emblematic of reactions in Canada to Darwin's theory of evolution during the final decades of the nineteenth century. But this assumption would be far too hasty. In fact, what is most conspicuous about the debate over Darwinism in Canada is its relative absence. Although Canadians discussed evolutionary theory on occasion, they did not engage in the rancorous feuding over Darwinism and religion that bubbled up in other places. Passing comment on what he calls Canada's "ambiguous encounter" with evolution, Michael Gauvreau observes: "In religious circles in Britain and America, Darwin's theory opened a great debate that raged for nearly three decades. By contrast, the Presbyterian and Methodist churches of English Canada did not have a single major contributor to the transatlantic discussion."[8] As for scientific engagements, Carl Berger is perplexed over the "puzzling reticence on the idea of evolution" in Canada, while Suzanne Zeller observes that "naturalists in British North America generally remained unusually taciturn on the subject of Darwin. No strong public defender of the theory emerged there during the half-century after 1859."[9] Indeed, a rather counterintuitive pattern of Darwinian sentiment emerges which discloses at best lukewarm adoption, at worst direct opposition. More typically, scientific practitioners exhibited a studied indifference to Darwin, while evolution received a rather warmer embrace by a significant number of denominational leaders, particularly among those touched by the heritage of Scottish theological culture.

In order to get a sense of these complexities, I first consider how Darwinism was read by a number of prominent scientific men in Victorian Canada—and Toronto in particular—in hopes of ascertaining the lenses through which they inspected the theory of evolution by natural selection. Next, I focus more single-mindedly on the situation at Knox College in Toronto, where theological commentators, drawing inspiration from their Scots Presbyterian counterparts, constructed their own modus vivendi with Darwin's challenging proposals.[10] The particularities of intellectual geography, it turns out, are critically important in grasping the dynamics of their encounter with Darwin. For, in contrast to the Belfast debacle, a notable equipoise over Darwinism—more reminiscent of Edinburgh—is discernible among the Knox College fraternity in Toronto.

Darwin and the Naturalists

In taking the kind of potshots at the theory of evolution that appeared in the newspaper press from time to time, populist anti-Darwinians in Canada could call on the august authority of Sir John William Dawson (1820–1899), the Nova Scotia–born geologist and prominent lay Presbyterian who became principal of McGill University in 1855. A graduate of the University of Edinburgh, where he studied geology with Darwin's old natural history teacher Robert Jameson, Dawson was elected to the Royal Society of London, authored many scientific and popular religious works, was president of both the British and American Associations for the Advancement of Science, and served as first president of the Royal Society of Canada. Dawson early on absorbed the Baconian philosophy of induction, which characterized a good deal of Scottish natural philosophy at the time, and mobilized it in conjunction with natural theology to launch his attack on

Darwin's theory of evolution. Indeed, he emerged as one of Darwin's most astute and articulate opponents and in this respect, if in no other, as successor to Louis Agassiz. For while Agassiz's creationism, with its idea of multiple centers of zoological genesis, was domiciled in the framework of a Unitarian-inclined *Naturphilosophie,* Dawson's theological pedigree was Scots Presbyterian. Indeed, a decade or so after his unsuccessful candidacy for the principalship of Edinburgh University in the late 1860s, his services were eagerly sought by the College of New Jersey (subsequently Princeton University) and the Princeton Theological Seminary—the heartland of American Presbyterianism. Over the years, he featured as a regular contributor to the influential *Princeton Review.*

Dawson's friend and confidant Charles Lyell had sent him a copy of *The Origin of Species* shortly after its appearance, and in April 1860 he penned a reply to Lyell, commending Darwin's empirical work on variation. In the letter, he placed welcome constraints on those far too inclined to multiply species upon species—not least Agassiz, who elaborated, seemingly ad infinitum, on the special creation of numerous different human races. But there his approval ended. Darwin's ideas on species and species change were simply "dreadful" and would do "much harm." And the theory's possibilities for "tempting generalization" were deeply troubling.[11] Reliable, hard-working, empirical naturalists could easily be enticed off course by the fatal charms of Darwin's alluring conjectures. And so, in his published review of *The Origin of Species* for the *Canadian Naturalist and Geologist* that same month, Dawson took twenty pages to expose what he considered to be Darwin's false methodology—one that sacrificed the "careful induction" of hard facts to "wild and fanciful" speculation.[12] Nearly half of Dawson's analysis was given over to lengthy citations from Darwin's work as he systematically sought to dismantle every key component in the Darwinian edifice. Darwin's comparison of natural selection to the calculated activities of pigeon breeders and the like was fatally flawed. For all the power of the breeder's dexterous hand, pigeons still remained pigeons, and it was simply nonsense to think that nature could "do still greater marvels" and generate new species. The "struggle for existence" was dismissed as mere fancy; breeders "cared for and pampered" their varieties, but Nature was supposed to improve "her breeds by putting them through a course of toil and starvation." No. To Dawson, that was all wrong. Nature was not so cruel, so ruthless, so tyrannical; beauty and harmony, not pitiless brutality, were her distinguishing hallmarks. Relentless, Dawson variously castigated Darwin's "loose way" of treating subjects, his capacity to lead "the unwary reader" astray, his tendency to lose "himself in the mazes in which he . . . continues to wander," and his "entire confusion of ideas" on the influence of environment. It all boiled down to a sorry case of a talented naturalist taking genuine pearls of induction but stringing "them upon a thread of loose and faulty argument" and employing "them to deck the faded form of the transmutation theory of Lamarck." No doubt, Dawson concluded, Darwin's theory would turn out to be a nine day's wonder, for he had to admit that fads did tend to catch on for a while. There were always those, not least young naturalists, who were only too "willing to adopt any amount of error rather than appear not to be on a level with the latest scientific novelties." Here then was Dawson's grievance; Darwin had violated the sound principles of Baconian induction and had too whole-heartedly joined forces with that breed of scientific "adventurers" who, like mythical suicidal lemmings, rushed headlong "into an unknown and fathomless abyss." What a tragic departure from "true science," which, Dawson insisted, "is always humble."[13]

Dawson's concerns did not go away. In 1872 he told the readers of the *Canadian Monthly* that if the Darwinian method were to triumph, "our old Baconian mode of viewing nature will be quite reversed."[14] The following year in a letter to Asa Gray, he expressed his alarm at the thought that any society intoxicated with "the doctrine of the struggle for existence" would "cease to be human in any ethical sense."[15] In *The Story of the Earth and Man* (1873)—a text that went through many editions in the years that followed—Dawson addressed the question of "Primitive Man" at the end of his account of earth history. He turned to Darwin's claims about human evolution and characteristically fastened on the absence of direct paleontological evidence in support of evolutionary transformism. "Even in respect to the question of species," he insisted, "in all the long chain between the Ascidian [sea squirt] and the man, he has certainly not established one link; and in the very last change, that from the ape-like ancestor, he equally fails to satisfy us on matters so trivial as the loss of the hair . . . and on matters so weighty as the dawn of human reason and conscience." His conclusion was curt: "We thus see that evolution as an hypothesis has no basis in experience or in scientific fact, and that its imagined series of transmutations has breaks which cannot be filled."[16] As for applying the survival principle to human origins, that maneuver was "nothing less than the basest and most horrible superstition"; the horror was too great to bear, for it made humanity "not merely carnal, but devilish" by reducing the crown of creation to the "lowest appetites" and ugliest impulses.[17]

For all Dawson's self-assurance, he knew deep down that he was already swimming against the tide. Irritated, he quipped that "it is characteristic of evolutionists to deny the intelligence of those who differ from them."[18] Feeling isolated and spurned, he nevertheless persisted. A decade or so later, when he brought out *Some Salient Points in the Science of the Earth*, he continued to call attention to the imperfect nature of the geological record and thus to evolution's inductive fragility. "The truth is," he maintained, "that such hypotheses are at present premature and that we require to have larger collections of facts." Philosophically, he went on, "all theories of evolution . . . are fundamentally defective in being too partial in their character" and far too inclined to make "unwarranted use . . . of analogy."[19] In championing this methodological stance, Dawson can be seen as rather typical of a Canadian scientific coterie "seeking support from a colonial society that was suspicious of the theoretical and ornamental and insistent on the directly useful." Canadian natural history, as Carl Berger tellingly observed, "blended scientific accuracy and a rigid factualism with a sense of wonder and a celebration of mystery."[20]

At least on the surface, Dawson tended to dwell on what he considered to be the scientific flaws—rather than the theological threats—in Darwinism, conspicuously trading on his geological expertise. Thus Michael Gauvreau claims that his critique "stemmed less from his religious or theological beliefs than from his scientific methodology."[21] To the extent that this assessment accords with Dawson's rhetorical stance, it very likely reflected the location in which he found himself—Montreal, a space very different from Toronto. Another Montreal resident, the distinguished but brusque Catholic naturalist Abbé Léon Provancher, who single-handedly launched and then controlled *Le Naturaliste Canadien,* made no bones about his religious objections to Darwin. In his assessment, evolution's scientific deficiencies were nothing to its theological absurdities and its conflict with established Catholic dogma. The journal (to which he contributed over six thousand pages of content, compared with less than one hundred from all the other contributors) was dominated by his own antievolutionary outlook and drew sporadic support from François-Xavier Burque.[22] The discussion often took on a political edge, not least when Provancher castigated Lamarckian transformism as "serving the self-pride of materialist French politicians" like the zoologist Paul Bert, the lawyer Jules Ferry, the lawyer and prime minister René Goblet, and the physician and later prime minister George Benjamin Clemenceau.[23]

Dawson, who had no desire to occupy either Provancher's theological or political terrain, felt more and more marginalized. Already alienated from his British colleagues on account of his failure to secure the principalship of Edinburgh University and the Royal Society's refusal to publish his 1870 Bakerian Lecture, he could only see the Baconian foundations of true science crumbling on every front. Committed to British empiricism and its inductive vision, Dawson had no more desire to be pigeonholed with speculative Darwinians than with Montreal's "prickly *nationaliste.*"[24] Thus marooned between materialism on the one hand

and ultramontanism on the one hand, he did all in his power to defend traditional British science from what he saw as both of these menaces. In identifying them as co-conspirators against scientific truth, Dawson certainly missed the opportunity to cultivate allies among Catholic anti-Darwinians in Quebec such as the geologist-priests Thomas-Étienne Hamel and Joseph-Clovis-Kemner Laflamme at Laval University.[25] But rather like the Ulster Presbyterians we have already encountered, Dawson had succumbed to what Zeller calls a "siege mentality," in this case as part of a Montreal "Anglophone community surrounded by an increasingly aggressive new ideology, ultramontanism, that dominated Quebec's largely Catholic society after 1840."[26] He thus devoted himself to preserving traditional Baconian science against both seditious metropolitan subversion and doctrinaire provincial dogma. He strongly felt a sense of duty to spread enlightened evangelicalism in his domestic province; as he explained to James McCosh and Charles Hodge, he could not leave Montreal for Princeton until "no hope remains that the Gospel and the light of knowledge can conquer our French Canada."[27]

Dawson's preoccupation with Darwin's brittle methodology, of course, did not mean that he had no religious stake in the Darwinian dilemmas, as he made clear when writing for other audiences. In articles for theological journals and popular science books for religious readers alike, he expressed his opposition to Darwin's theory. At least since the publication of *Archaia; or, Studies of the Cosmogony and Natural History of the Hebrew Scriptures* (1860), Dawson had resorted to the idea—earlier championed by Hugh Miller—that each of the creative days of Genesis involved a metaphorical allusion to lengthy geological epochs so that he could maintain the harmony of Genesis and geology.[28] Thirty-five years on, in 1895, when he drew together—under the title *Eden Lost and Won*—a series of articles he had penned on a range of subjects for the *Expositor*, a nonconformist scriptural magazine, he reiterated this concordist reading of the Mosaic record in opposition to the "aggressive forces of agnostic philosophy and destructive criticism" to which the Hebrew Bible was being subjected.[29] Indeed, in the interim—for example in his 1887 *Modern Science in Bible Lands*—he resorted to Augustine's conception of the *dies ineffabiles*—that is, the "days" prior to the creation of solar days—to provide further warrant for conceiving of the Mosaic days as "Divine Ages."[30] He assured the audiences for these works of evangelical apologetic that Genesis in fact anticipated the Nebular theory, that "Biblical history of the antediluvian time and of the Deluge will be more and more valued as knowledge advances," and that traditional doctrines like the fall from grace remained immune to subversion "even by the current theories of evolution, except insofar as these occupy the entirely irrational ground of agnostic causelessness."[31] Here, too, he panned the "disagreeable prospect" that humanity's first parents were either "no better than modern Australian savages" or the "trembling survivors of ice and cold, struggling for existence on the shores of an arctic sea" in favor of the Mosaic picture of Adam in the Garden of Eden.[32]

For readers of the *Princeton Review,* Dawson happily acknowledged that he himself was routinely "characterized as one of the few remaining naturalists who do not believe in the theory of evolution" and yet again insisted, in true Baconian fashion, that "We must have the facts first and then go on to the conclusions. We must be empirical first, afterward rational; gaining the materials of knowledge before arranging them in scientific formulae."[33] Within a few pages, he was predictably railing at those evolutionists "who insist on adapting all classification to imaginary theories of descent" and thereby were implicated in "sapping the foundations of science."[34] Not surprisingly, Ernst Haeckel and Herbert Spencer came under the whiplash of his tongue; the former's monism and the latter's agnosticism were "absolutely repugnant," serving only to "degrade rather than to elevate." Should their sordid speculations gain a foothold in the curriculum, "it would be necessary for the safety of humanity that natural science . . . should be abolished as a nuisance, or even as an unnatural crime."[35]

For all the staging and, indeed, self-staging of Dawson as antievolutionist extraordinaire, even in these works it is clear that, whatever his distaste for Darwinian natural selection and struggle, he did allow space for organic transformation of one sort or another. It bothered him that Darwin and Spencer should usurp the idea of "development" and by what he called "a scientific sleight of hand" confuse evolution as an active principle or force with evolution as "an actual process going on under ascertained laws and known forces."[36] If evolutionary progress were to be understood as the empirical outworkings of divine design, it could be different. So, even as he grounded his opposition to Darwinism in Baconian induction, he insisted that nature should be read teleologically. It was a recurring refrain. In 1873 he categorically declared that evolution was unacceptable because it "removes from the study of nature the ideas of final cause and purpose."[37] But as the years passed he began to conceive of evolutionary change in terms amenable to design.[38] In 1890, his comprehensive review *Modern Ideas of Evolution as Related to Revelation and Science* was published by the Religious Tract Society. In it, he exploited the tensions in current evolutionary thinking,[39] carefully discriminating between what he called monistic, agnostic and theistic evolution. "It is true that there may be a theistic form of evolution," he declared, "but let it be observed that this is essentially distinct from Darwinism or Neo-Lamarckianism . . . It necessarily admits design and final cause." Yet Dawson plainly felt uncomfortable with even this qualification, sensing some intrinsic "incongruity between the methods supposed by evolution and the principles of design, finality, and ethical purity." For "theistic evolutionists," there was always "the danger that in the constant flux of philosophic opinion they will find their system of theology, which at present rides so triumphantly on the flood-tide of a popular movement eventually stranded . . . on the sandbanks of the ebb." The wise course of action was clear: avoid commitment "to any of the current forms of evolution."[40] All were afflicted with such vices as circular reasoning, flimsy evidence, unproved hypotheses, dangerous ambiguity, and a slippage toward naturalism.

The allergy to speculation that troubled Dawson throughout the course of his scientific life was doubtless contracted through his exposure to Scottish Common Sense philosophy, which had long outlawed speculative hypothesizing in favor of down-to-earth Baconianism. The same intellectual conditions influenced the outlook of his close friend, the Scottish archaeologist Daniel Wilson, who spent nearly forty years at the University of Toronto, first as professor of history and English literature and later as president of the university. And yet Wilson's location in Toronto seemed to moderate his rhetoric, allowing him to exult, at least in certain circumstances, in the virtues of scientific conjecture. His situation was different, and the Montreal conditions that served to induce a sense of intellectual and spiritual isolation in Dawson were conspicuously absent in British Canada's first city. So in his presidential address to the Canadian Institute in January 1860, Wilson worked hard to find an honorable space for the local successes of Toronto practitioners in the pantheon of global science. Noting the practical achievements of Canadians in forest clearance, swamp drainage, and road construction, he was proud to record the role played by Toronto's Magnetical Observatory in speculations about the plurality of inhabited worlds. It was from "that little building which rears its modest tower in the University Park" that scientific advances were achieved which widened "the range of our knowledge, and also the area wherein fancy may freely speculate."[41] Against the backdrop of what he called "the direct commercial and utilitarian results of Canadian science," Wilson evidently wanted to stake Toronto's claim as an interlocutor in the most recent scientific conversations. In this, he echoed the earlier sentiments of Sir Henry Lefroy—military officer, earth scientist, superintendent of the observatory, and a founder of the Canadian Institute—who had "fought hard to convince Canadians to accept the torch of imperial science."[42] The observatory stood as a symbol of Canada's participation in British scientific culture and in the advancement of global science more generally. After all, the continuing maintenance of the observatory had long been justified on the grounds that it made vital contributions to the understanding, not just of "the Natural & Physical History of Canada, but of the whole World."[43]

The following year, in a second presidential address to the Canadian Institute, Wilson further expanded on the close ties between Canadian and British metropolitan science. Recalling with pleasure "the visit to our colony of the Heir apparent to the British Throne" and "the awakening of pure and lofty sentiments of patriotism and loyalty" that the event had induced, he exulted in "all the glories of that empire . . . which is now girdling the world with a glorious confederacy of provinces, alike united in freedom [and] in intellectual progress." Toronto thus took its place "in all the triumphs which mark the progress of Science, wheresoever achieved." And so with patriotic zeal verging on the jingoistic, Wilson rejoiced in the laying of the transatlantic telegraph cable: "The great pulse of the empire throbbed in sympathy with that of the proud young western national kindred with itself, and the common ancestral blood seemed to kindle anew into generous aspirations, with the consciousness that time and space had been annihilated, and the broad Atlantic no longer severed them and us from the vital heart of Britain's world-wide empire. Science had her triumph."[44] Thus, while Dawson was increasingly burdened with a sense of his marginalization from science's metropolitan core, Wilson was constantly scanning the horizon for signs of intellectual affiliation and scientific continuity with the colonial heartland. Dawson looked back to the glories of British science that once was; Wilson reveled in its contemporary conquests.

In this context, Wilson's preliminary appraisal of Darwin's *Origin of Species* was notable for its rhetorical distance from Dawson's evaluation. As he reviewed current thinking on the species question across a range of disciplines in 1860, Wilson turned to the "eminent English Naturalist" Charles Darwin and his account of organic variation under natural selection. To Wilson, the *Origin of Species* was "no product of a rash theorist, but the result of the patient observation and laborious experiments of a highly gifted naturalist, extending over a period of upwards of twenty years, and . . . it will be found to embody thoughts and facts of great permanent value, whatever be the final decision on its special propositioning." Certainly, Wilson had his doubts—and the following year he elaborated on them in rather more detail—but the tenor and tone of his commentaries markedly contrast with Dawson's assault on Darwin as an outrageous speculator. The "novel and highly suggestive views on the origin of species by means of natural selection," Wilson reported, could not be treated "with too sincere respect even while rejecting them. They are no rash and hastily formed fancies of a shallow theorist, but the earnest convictions of an eminent English naturalist of great and varied experience." Moreover, they had been "heralded by the favourable testimony of some of the most cautious and discriminating among his scientific contemporaries."[45]

In the end, however, Wilson could not surrender to the forces of the Darwinian empire, predominantly on account of its troubling implications for the human species. He always cast the species question into the arena of human variation—whether to rebut Agassiz's racial polygenism or to confirm humanity's psychic unity. So it was not surprising that in 1861 he immediately fastened upon "the bold naturalist's views on the origin of man himself"—even though, he noted, Darwin had merely hinted that "light will be thrown on the origin of man and his history."[46] The intellectual atmosphere was already thick with conjectures on such subjects, and Wilson simply threw Darwin into the context of the anthropological speculations of figures like Johann Friedrich Blumenbach, Charles Pickering, Bory de St. Vincent, Samuel George Morton, George R. Gliddon, and Josiah C. Nott, all of whom in one way or another obsessed over racial classification.[47] Adrian Desmond and James Moore have recently unpicked the racial fabric of Darwin's entire schema, urging that issues of race, slavery, and abolition constituted the "sacred cause" to which he was all along devoted. Pigeon breeding, they point out, was a favorite analogical hunting ground among the cognoscenti tracking human racial lineage.[48] That was certainly so for Wilson. He eagerly fastened on Darwin's "favourite illustration, the domestic pigeon" to confirm that its persistence as a species—whatever varieties had emerged—was "to a far higher degree characteristic of man." To Wilson, climatic conditions, the influence of domestication, the forces of artificial civilization, and the like, were every bit as active in human variation and conspired to "yield still stronger proofs that the man of Europe, of Egypt, and of India, are alike descended of one primal stock."[49] In thinking about Darwinism, Wilson netted together pigeons and people.

Wilson had long abhorred polygenetic rationalizations of racial servitude; like Darwin, he came from a resolutely antislavery family. His family was Baptist and melded evangelicalism with social concern. He had grown up in Edinburgh where, prior to taking up his position at the University of Toronto, he had worked as an engraver, literary hack, and watercolorist. Although he became an Anglican, he retained his evangelical convictions, abhorring what he regarded as the empty ritualism of the Tractarians and the high church extremism of the Oxford Movement.[50] He retained, too, his loathing of racial hierarchy. When he arrived in Philadelphia in 1853, he was horrified to hear that some people denied that all races had sprung from the same stock. But his interest in such matters was not simply a product of his social piety; it was a subject in which he had much scholarly investment. Wilson had long experience in Scottish archaeology and adopted the Danish three-fold system of periodization, which divided human prehistory into stone, bronze, and iron ages. Researches on archaeological relics had led him to examine skeletal remains and engage in the measurement of human skulls as part of a project to elucidate the early settlement history of Scotland. His published work, such as *The Archaeology and Prehistoric Annals of Scotland* (1851) and later the two-volume *Prehistoric Man* (1862), confirmed his reputation as a significant scholar and was instrumental in establishing prehistory as a serious science in Britain.[51] In the New World, his interests in cranial types mushroomed, and he threw himself vigorously into the debate on racial origins, siding with the monogenists against the polygenists. For Wilson, the unity of the human species, together with his conviction that European prehistory could be illuminated by attending to North America's indigenous native culture, was an abiding conviction, and his empirical work on everything from metallurgy to ceramics was geared to confirming humanity's common constitution. It was an old Scottish theme—Common Sense philosophy—which gathered all humanity into a single human sensibility. This Scottish Enlightenment *mentalité*, as Bruce Trigger points out, combined with "official Christian dogma" to confirm "that all groups of human beings shared a common origin, as a result of which they had a similar nature and were equally capable of benefiting from intellectual progress."[52] This principle enabled Wilson to find links between peoples temporally and spatially far apart—such as the funeral rites of ancient Saxons and contemporary Omaha Indians—and this led him to the conclusion that "man in all ages and in both hemispheres is the same; . . . amid the darkest shadows of Pagan night, he still reveals the strivings of his nature after that immortality, wherein also he dimly recognizes a state of retribution."[53]

Wilson had already expressed himself in print on such issues prior to the publication of *The Origin of Species*. In 1855, for example, he announced that biogeographical inquiry would confirm the biblical account of human descent from an original couple.[54] Shortly after it appeared, his own *Prehistoric Man* (1862) revealed the extent to which he was revolted by polygenism. To him, human differences were literally only skin deep and could be explained perfectly well by human intermixing, environmental adaptation, and culture contact. Within the confines of a biblical cosmogony fashioned along the lines of Hugh Miller's concordism, Wilson championed the global unity of the human constitution by referring to the faculty of speech, structural grammar, tool-making practices, funerary rites, architectural forms, and so on. Using America's "living present" to throw light on "Europe's ancient past," he brought "the civilized man and the savage . . . face to face" to show beyond all doubt that God "hath made of one blood all nations of men." Indeed, America's native peoples, he insisted, carried within them "the germs of all later triumphs of chemistry, electricity, mechanics." All this industry, designed to unveil "man's innate capacity" and "what is common to the race," confirmed his earlier attacks on the polygenetic claims of Morton, whose work he had read before he left Britain, as well as Gliddon and Nott's.[55] Wilson had no time for the thought that there were such things as pure, uniform racial groups, and he left no stone unturned. He carefully sifted through their "evidence," recalculated their cranial measurements, and revised their data to come to the precisely opposite conclusion.[56] Three decades later, a collection of ethnographic essays posthumously published in 1892 showed his persistent interest in issues revolving around human hybridity, heredity, and intermediate types and his determination to undermine the popular presumption that brain weight and size were any "measure of intellectual power."[57]

Wilson's Scottish credo, anchored to monogenist moorings, thus profoundly shaped his encounter with Darwin. His *Prehistoric Man* made its appearance at precisely the time when Darwin was shifting the whole framework of debate from single or multiple creation to evolutionary transformism. But that philosophical heritage meant that Wilson cast evolution so unmistakably into a racial framework that he focused right from the start on its implications for humanity. In *Prehistoric Man*, in his only discussion of Darwin's "ingenious argument" in the two volumes, he turned for a second time to pigeon breeding and what it meant for understanding human diversity. To Wilson, Darwin simply had not succeeded "in tracing the slightest indications of that favourite illustration of the instability of species, the pigeon, being developed out of any essentially distinct form." He had done no more than show how varieties within a species could be produced through domestication and environmental adaptation. It was precisely the same for humanity. Pigeon variability furnished "interesting analogies readily applicable to the so-called races of men."[58]

Scottish philosophy also predisposed Wilson to exempt human consciousness from the operations of reductionist natural law and encouraged him to welcome Alfred Russel Wallace to the campus of the University of Toronto in the late 1880s. Wallace had come to believe, particularly after 1870, that natural selection was impotent to explain the development of the human mind, and he thus exempted it from the very law of natural selection of which he is often considered to be the codiscoverer.[59] For his part, Wilson was certain that humanity's moral and cognitive faculties were fundamentally different from mechanical animal instinct, and he thus repudiated "the monstrous notion" that any form of evolutionary psychology could "account for the origin of the intellect and living soul of man by development." To think that natural selection could explain human consciousness was to get things the wrong way round; it was, rather, that human "intellectual power" explained why "man . . . has triumphed over all other animals."[60]

Wilson certainly was no wholehearted convert to Darwin. The legacy of Scottish Baconianism was simply too deeply ingrained for that. And yet the rhetorical stance he adopted in dealing with Darwin's challenging intervention was markedly different from that of his good friend Dawson stranded in Montreal. Wilson's judgments were expressed in the muted and modulated tones of a Toronto Scotsman at the intellectual heart of British Canada. Neither religious protectionism nor cultural isolation were part of his mindset, as a letter to Bishop Isaac Hellmuth dated 4 May 1877 illustrates: "I need not assure you of my sympathy in reference to all that pertains to the clear setting forth of evangelical truth . . . But I must also inform you that I no less strongly desire to see the untrammelled freedom of scientific and philosophical research. Truth has nothing to fear in the long run from the researches of such men as Darwin and Huxley."[61]

Bacon's bequest was no less conspicuous in the judgments of the geologist Edward John Chapman (1821–1904), a civil engineer who had come from University College London in 1853 to take up the first professorship of mineralogy and geology at the University of Toronto. Like Wilson, he was critical of Darwin's theory but, according to Zeller, was equally convinced that "Dawson's entrenched position went too far."[62] When he came to review Dawson's *Archaia: Studies of the Cosmogony and Natural History of the Hebrew Scriptures* in 1860, for example, he confessed that he could not "go with the author to the full extent of his argument."[63] Chapman had been educated in France and Germany and was the author of a number of technical works on mineral identification, as well as a couple of volumes of poetry.[64] Ordinarily chary of theoretical discussion, he was most at home in conveying geological knowledge to general readers or producing guides to help farmers identify minerals. In 1860, however, he announced his views on the newly published *Origin of Species* to the readers of the *Canadian Journal*. Still, it was at the bar of facts, facts, and yet more facts that Chapman judged Darwin's offering. His language was complimentary, his stance condoning. Darwin's book, he confirmed, displayed its author's "laborious collection of facts and his skilful deduction," and even though readers might not be convinced by his theory of evolution by natural selection, Chapman was sure that "no one . . . can lay down Mr Darwin's book, so remarkable in many points of view, without feeling that a large accession of new thought has been added by it to our common store." Besides, Chapman fully understood why any serious naturalist would hesitate over the view that attributed each and every species to a separate act of creation. In an age of excessive taxonomic splitting, when varieties were too frequently elevated to the status of species, anyone who had undertaken a "close contemplation of Nature" had been visited by "sundry hauntings" along the lines of Darwin's conjecture. The transitional stages of fetal development, the existence of rudimentary organs, and the homologies of organic structure were remarkably perplexing if indeed it were the case that "all species were separate and distinct creations."[65]

Nevertheless Chapman did not find Darwin's solution convincing. Despite Darwin's "ingenious and eloquent reasonings," his theory "offered no real help in our difficulty." Chapman was troubled by the slipperiness of Darwin's understanding of "species," the absence of any serious "true transition-link," and what he believed were critical similarities between his project and Chambers's *Vestiges of the Natural History of Creation*—despite Darwin's protestations to the contrary. Again and again he stressed the empirical fragility of the theory, Darwin's over-extended inferences, his resort to "blind and gratuitous surmises," his inclination to indulge in special pleading, and his assuming the validity of "data altogether denied by the greater number of our most eminent geologists." None of this meant that Chapman denied the reality of natural selection as a mechanism however. It was "undoubtedly a modifying power or principle of recognised action," though suggesting that it could induce "generic changes" grossly overestimated its power. "Let us do this theory no injustice," he announced. "It certainly does afford a rational explanation of the remarkable facts detailed above; but when tested by other facts, it fails entirely. It is comparatively easy to invent a theory in explanation of a particular series of phenomena, provided we be allowed to exclude all collateral facts from consideration." At bottom this was Chapman's concern: Darwinism represented the triumph of theory over evidence, of supposition over information, of fancy over fact. And so Chapman steadfastly held to the older, natural theology tradition. "We are asked how the world came to be peopled by so very many different plants and animals," he reflected. "We reply, by the act of the Creator." Whatever the difficulties posed by rudimentary organs, homological echoes, fetal transformism and so on, they were much more satisfactorily explained "as parts of a great plan, conceived and carried out by the Almighty in his wisdom."[66] Still, Chapman did increasingly surrender ground to Darwin and conceded in an 1863 review of Lyell's *Geological Evidences of the Antiquity of Man* that his theory had "some strong claims to consideration."[67]

In key respects, Chapman's assessment of Darwinism paralleled that of the Irishman William Hincks (1793?–1871) who had come from Queen's College, Cork, to the University of Toronto in 1853 when he was in his late fifties, defeating in the process the brilliant young Thomas Henry Huxley and John Tyndall, who had also been applicants for the chair in natural history. The fact that Hincks's younger brother Francis was currently premier of the province of Canada probably did not harm William's candidature.[68] At the same time, as Brian McKillop astutely observes, "the authorities who had hired William Hincks had many occasions to remind themselves that they had made a wise choice . . . [for he] gave every indication that he was 'enlightened' in exactly the fashion the hiring authorities had expected him to be."[69] William Hincks was an ordained Unitarian clergyman who had studied at Manchester College, York, and a devoted supporter of Joseph Priestley. He moved increasingly toward a career in natural history after holding the editorship of the leading Unitarian journal, the *Inquirer*, during the 1840s. He had continued to champion and revise in fairly far-reaching ways the earlier system of classification known as Quinarianism that had been put forward in the 1820s and 1830s by William Sharp Macleay and William Swainson.[70] This was a taxonomic scheme that divided the animal kingdom into five subkingdoms, each of which was split into five classes, then five orders, and so on. Hincks was a belated apologist for a modified version of this idealist morphological system, and he put it to work in his own museum exhibitions. As a consequence, Jennifer Coggon remarks, "quinarianism stirred like a phoenix rising from the dust of Toronto's museum shelves."[71]

Like other Canadian naturalists, Hincks was skeptical about Darwin's theory and focused on the issue of the reality of species over against nominalist definitions. He made this especially clear in an 1863 review of a course of lectures that Huxley had delivered to workingmen on the origin of species. Hincks began by declaring that the "grave questions" that the lecture series raised forced him "to express our reasons for not assenting to the hypothesis defended." He was worried, for example, by Huxley's willingness to subject an audience of workingmen to "speculative views on the most recondite question his science afforded" rather than sticking to "the established principles and interesting facts of natural science." Given his convictions about ideal plans in nature, Hincks challenged Darwin's presumption that species could transcend fixed boundaries, as well as his use of the domestic pigeon analogy. He was convinced that transmutationists were just far too willing to sacrifice the stable principle of species identity on the insecure altar of prior theoretical conviction. To be sure, some developments in organisms were empirically discernible, but these always occurred within definite, predetermined confines. As he put it, species were "fixed in the nature of things and only liable to modification by external causes within certain limits." Until species fluctuation should actually be "observed" to take place, it was "not the business of the philosophical inquirer to form some theory respecting the origins of the various species."[72] His own nested typological system was preferable, disclosing as it did the interdependence of organic communities; the equilibrium of the natural order attested to the wisdom of the Creator, the master-builder who had designed nature's typical forms. In other hands, the idea of transcendental plans could be mobilized to produce a creative encounter with evolution—as in the case of James McCosh, for example—since it could facilitate the cultivation of a natural theology uncoupled from William Paley's classical version.[73] But for Hincks, it confirmed his fixist conception of the plan of nature and the position of species within that predetermined scheme of things.

In many respects, Hincks may be seen as emblematic of the final stirrings of a tradition of natural history harking back to earlier systems of taxonomy and organization. On his death in 1871, his chair was occupied for a short period by H. Alleyne Nicholson, who returned to Britain to take up a professorship at the University of Durham after only three years. Another Scottish-trained natural historian, Robert Ramsay Wright, assumed the position in 1874. In marked contrast to Hincks, Wright conspicuously disentangled science from religion in his engagement with Darwinian evolution. During the 1870s and 1880s, he praised the unrivaled richness of *The Origin of Species* and began constructing a program of research inquiry conceived on evolutionary principles.[74]

Nowhere, perhaps, was Wright's settled assurance about the truth of Darwinian biology more plainly advertised than in his *Introduction to Zoology* (1889), a text intended for use in schools. Here he simply recorded what he called "the generally accepted view" that the geological strata disclosed *"descent with modification."*[75] This meant, he told his youthful readers, that when "one generation exceeds the previous one . . . there will soon be a competition or struggle between the individuals for suitable food, and that the strongest and best adapted to survive will be those, on the whole, that do survive. Along with this," he went on, "we have to take into account the tendency of organisms to vary and to transmit their peculiarities to their descendants. Those individuals whose organs vary in such a way as to adapt them better, however little, to their special circumstances of life, are, therefore, rendered better fitted to survive." It was Darwinian natural selection pure and simple. And just so that there would be no doubt, Wright concluded, "The theory stated in the preceding paragraphs is that of the Origin of Species by Natural Selection, associated with the names of Darwin and Wallace."[76] To be sure, Wright acknowledged that contemporary biologists differed in their account of the cause of variation, some attributing it to environmental factors, others to internal forces. He himself sided with the American neo-

Lamarckian evolutionists, particularly Edward Drinker Cope who allocated a role to changed habit and the direct effect of environment in organic modification, over against what he called the "strict Darwinists."[77] But there he left the subject, content to report what he took to be the current biological orthodoxy of his day.

Other scientific voices explicitly espousing Darwinian evolution in the Toronto orbit could doubtless be sought. But they were few and far between. This paucity was simply because overt engagements of any stripe with evolutionary theory—whether expressing vigorous disagreement, half-hearted support, or wholesale endorsement—were mostly conspicuous by their absence in a scientific culture dominated by pragmatic instrumentalism and an antipathy to theoretical speculation. According to Carl Berger, the rendezvous with Darwin in Canadian scientific circles overall "remained restrained and muted." In Berger's view, "Canadian naturalists were reluctant to be drawn into theoretical discussion because Darwin's hypothetical mode of argument was so alien to the tradition in which they had been trained and also because his theory did not impinge directly upon the collecting and classifying activities in which most of them were engaged."[78] In large part this stance was born of the Canadian love affair with a Scottish tradition that valorized the Baconian ideal. For when Canadian naturalists felt the need to make explicit the philosophical foundations on which their enterprise was built, they routinely resorted to Bacon's insistence that genuine understanding was bound up with the "patient accumulation of detail, not with deductions from supposedly universal principles." In practice, this meant that the bulk of scientific endeavor rotated around the collecting of information about plants and animals and the amassing of samples and specimens—crucial scientific pursuits in a region where "little was known about the natural history of even long-settled districts."[79]

Inventory science, as this impulse has been branded, thus resonated with a related Baconian virtue that perfectly suited the needs of a new nation in search of itself—utility. The incremental impulse of Bacon-style natural science was well equipped to deliver solutions to the practical problems settlers faced in their encounter with a harsh environment characterized by climatic extremes, primeval forest, and the unyielding reality of the cold, hard Canadian shield. For it presented useful ways of identifying good soils for farming, valuable mineral deposits for industry, and plants with commercial potential. Derived in part from the Baconian legacy of the Scottish Enlightenment, this cataloguing tradition, as Zeller notes, "lent a sense of purpose and meaning to the arduous task of settling British North America" and thereby cemented the ties between scientific endeavor and nation building.[80] In this context, good science just *was* about the slow, patient, accumulation of information about the new transcontinental nation; it was about systematically surveying a vast terrain and evaluating its potential to provide for the needs of its inhabitants. Theoretical conjecture and speculative inference were distracting luxuries that could not be compared with the "technological signs of material progress and economic development—canals, railways and electric telegraphs."[81]

Knox, Bacon, and Darwin

In this scientific context it is instructive now to turn to another Toronto location whose conceptual architecture was also erected foursquare on Scottish foundations and where Darwinian evolution was the subject of serious conversation during the later decades of the nineteenth century—the Presbyterian Knox College. The college had been established in 1844, in the wake of the formation of the Free Church of Scotland, with which Knox's founders were in deep sympathy. It received its charter from the government of Ontario as a degree-granting institution in 1858. Various ecclesiastical amalgamations later, it sought and was granted affiliation with the University of Toronto in the mid-1880s. Throughout its first half century, the college's culture was shaped by Scottish philosophy, its faculty dominated by Scottish-trained professors, and its curriculum fashioned largely on the model of Edinburgh's New College. The orthodox confessional culture that these forces coalesced to produce had developed, Brian Fraser observes, "in the midst of the evangelical-moderate debates in Scotland in the late eighteenth century and forged in the controversies surrounding the Disruption in the first half of the nineteenth century."[82] So how did this theological community map a path through the Darwinian challenges that are routinely considered to have traumatized Victorian religious believers?

Because Knox College published its own monthly magazine, we have a rather useful entry point to the way Darwinism was handled within a denominational community profoundly shaped by the same Scottish Common Sense philosophy that dominated the intellectual atmosphere of British Canada more generally. The *Knox College Monthly* was launched in 1882 under the editorship of James Ballantyne, who later would return to the college as professor of church history in 1896. James A. MacDonald took over a couple of years later in 1885, and soon the periodical received high praise from a range of eminent theologians, securing in the meantime a widespread readership across Canada. It promised to deal directly with tough issues, especially in the face of contemporary challenges from a variety of quarters, and its editorial manifesto specifically pointed to Darwinian evolution as a theory that commanded the attention of every serious Christian believer.[83] The journal's leadership kept a close eye on theological and ecclesiastical developments within the Free Church in late Victorian Scotland and sought to import into their own Canadian context the perspective of its intellectual leadership—one which was "less-strait-jacketed, less censorious, more broadminded" than hitherto.[84] What is noticeable, then, about the way in which contributors to the journal approached Darwinian matters is the rhetorical tone—ameliorative, accommodating, and bereft of the animosity that characterized the reaction of their counterparts in other places. It stood in marked contrast, too, to the anxiety exhibited by the Toronto Baptist College, which rejected integration with the newly federated University of Toronto in the mid-1880s at least partly on account of serious concerns about the "prospect of leaving the teaching of biology to the university professoriate." At any rate, that was how Daniel Wilson saw it, for he scribbled in his diary for July 1884 that the Baptists wanted scientific teaching to "be put under a theological censorship fully as admirable as that of the Dominicans in Florence in Galileo's time."[85]

In the main, *Knox Monthly* writers worked hard to absorb the essentials of the new theory of evolution and even to rethink certain doctrinal principles in its light. Acrimonious opposition, literalistic readings of the Hebrew Bible, and resolute adherence to a fixist creation were noticeably absent. On matters of interpretation, for example, G. M. Milligan made it clear in his 1886 reflections on the Mosaic cosmogony that "long before the sciences of Geology and Chemistry came into existence . . . discerning minds [were] puzzled with the contents of the first chapter of Genesis."[86] Origen and Augustine, he pointed out, had been bewildered by some of its claims and had labored long and hard to find coherent ways of making sense of the text; they certainly had not advocated any literalist reading. Milligan's purpose in revisiting the early history of exegesis was to resist the terms of debate in which Huxley had recently engaged William Ewart Gladstone, four times prime minister of the United Kingdom, in the pages of the *Nineteenth Century* over the Hebrew cosmogony.[87] Huxley, Milligan's readers were told, "alleges that in no way can the contents of the Mosaic account of the Creation be made to harmonize with the teachings of science."[88] Milligan disagreed, favoring a concordist reading of the text, but he pressed home the point that Huxley was far too inclined to interpret with unwarranted precision very general Hebrew terms for certain creatures for the express purpose of declaring the Genesis text to be forever incompatible with modern scientific findings. Besides, there was a history to biblical interpretation—just like explanations of nature—and he found it surprising that Huxley could happily discount "Cuverian interpretations which are now fifty years old" while denying any comparable revisionism to biblical exegetes. "Fixity and finality," Milligan concluded, "must not be claimed for, or fastened upon, either the students of Nature or Revelation."[89]

The mindset that Milligan displayed in his account of the Huxley-Gladstone altercation exemplified the spirit of the *Knox College Monthly*'s engagement with Darwinism. One of the strategies that several contributors deployed in assessing evolutionary theory was to capitalize on those structural aspects of Darwin's account which most obviously reverberated with architectural echoes of William Paley's natural theology. As John Brooke has pointed out, such critical concepts as the adaptation of organisms to their environments and the idea of the harmony of nature were as central to Paley's cosmos as they were to Darwin's.[90] As for the Knox College community, Nina Reid perceptively notes that its theologians "were building upon the deeply engrained assumptions of natural theology which found their way into *On the Origin of Species.*"[91] The issue of design routinely surfaced in the magazine, often with the aim of teleologizing Darwin . . . and evolutionizing Paley. Tackling the subject of "Evolution and the Church," a Knox College alumnus, William Hunter, argued that evolutionary teleology far outstripped its Paleyite counterpart. Design by wholesale, as it were, was grander than design by retail! Having reviewed the standard argument that discovering a watch on the heath could reasonably lead to the conclusion that it had a watchmaker—a designer—he went on:

> But suppose, instead, the man was taken to the factory and saw watches made by machinery . . . If it be an argument from design that a man can make a watch, is it not a sublimer argument that there is a man who can create a factory which turns out thousands of watches? If it be an evidence of design that God adapted one animal to its place and function, is it not greater evidence of design if there be a system of such adaptations going on from the beginning? Is not the creator of a system a more sublime designer than the creation of any single article? . . . [Evolution's] principle of variation and of natural selection simply show the existence of an *internal principle of transformation,* and thereupon the idea of design resumes its whole empire.[92]

Nothing in Hunter's analysis, of course, meant that Darwin himself considered natural selection a teleological operation—that was something on which he wavered and which he debated at length with the Harvard botanist Asa Gray.[93] The point is that it shows Hunter's efforts to take seriously what he called the "plasticity of living forms," to use nature's malleability under the operations of natural selection to recast Paley's scheme in evolutionary form, and to make clear that design was as compatible with gradualism as with fiat.[94] His motivations sprang from his concern that adversaries of Darwin tenaciously clung "with blind affection to the old ways"; their "hostility to every form of evolution" meant that they had "shut their eyes to whatever truth it might contain." Such opponents were too inclined to content themselves with lampooning evolution as the view "that your grandfather was an ape." Their strategy was unwise and unjust. "When we wish to turn our guns against a theory," he insisted, "we must not take the testimony of its enemies, nor of its too zealous friends, as to its meaning." For himself, far from "obliterating evidences of design," he was sure that evolution only succeeded in lifting "the teleological argument to a higher plane"[95]—a vision he supported by calling as witness such luminaries as Asa Gray, James McCosh, Joseph LeConte, James Dana, and Lyman Abbott.

Nor was Hunter a lone voice. A decade earlier, William Dewar, a Knox student, had turned his readers' attention to the subject of biology and theology. Recognizing that "the Biology of today is inseparable from the theory of evolution," Dewar insisted that its truth or falsity was "purely a scientific question." It was used by the biologist "as a 'working hypothesis,' exactly in the same way that the chemist uses his atomic theory," and so he was happy to acknowledge that "evolution is being every day put to the severest test, and will eventually be confirmed or rejected finally, according as it satisfies or falls short of the facts of the science."[96] As for its religious implications, he claimed to find nothing in "the laws of Heredity, Variation and Natural Selection" which could not be reconciled with the early chapters of Genesis rightly understood. As for the question of design, he told his readers that "if evolution be true, it is a law, a method of operation, and accordingly it itself becomes the embodiment of design."[97]

What is most apparent here, in contrast to many of the judgments of the Toronto scientific community, is the accommodating language that interlocutors used in their encounters with Darwin. Plainly a less mechanistic, more organicist, rendering of Paley went a long way toward facilitating a creative rendezvous with evolution's chief proponent. But what is also notable is the extent to which various philosophical assessments of teleology in the pages of the *Knox Monthly* also conceded just how conceptually fragile the argument from design really was. Alexander Blair—another Knox student—expressed the long-standing Scottish Calvinist doubt about the theological propriety and intellectual potency of design arguments in general when he candidly declared: "The simple matter of fact is that a logical argument cannot possibly be constructed from these facts which will prove the existence of a Deity. We cannot have more in the conclusion than we have in the premises."[98] To Blair, the argument was not entirely impotent, but its force was far from delivering a logically compelling proof. Indeed, Blair expressed a remarkable sensitivity to the experiential dimensions of the idea of nature's beauty and to those factors that moderated its existential force. "Those who live in regions of romantic scenery, where the landscape is of surpassing beauty, and all nature wears the garb of sublimity and is clad in robes of majesty," he remarked, "from their very familiarity with such scenes take no special notice of them."[99] Any logical power the argument from design had was necessarily modulated by psychological perception and local particularity. F. R. Beattie also contributed a sequence of pieces on the design argument during the winter of 1885–1886 in which he acknowledged a range of philosophical problems with its standard formulation. There were the troubling concerns stemming from Kant's suspicion that "the notion of finality is merely *subjective* and regulative in our thinking, and not *objective* and constitutive in nature," but there was also the simple fact that, even at its best, the argument could establish no more than "the existence of an intelligence" above nature—a far-distant call from either an "*infinite* intelligence" or the Creator disclosed in the Hebrew Bible.[100] To be sure, this did not mean that the teleological argument was of no value; it simply meant that it had to be used with circumspection. Staking too much on the issue of design in judging Darwin's theory was altogether wrong-headed.

Even on matters of human origins, there is a striking absence of hostility. William Hunter, for example, did not shirk the implications of Darwinism for human descent. "Man is not less a work of art, because he is gradually formed," he insisted. "Indeed the 'ascent of man' is quite consistent with the universal law that matter must be elaborated and specially prepared before it can be the organ through which a higher power is manifested." Not surprisingly, he drew inspiration from the efforts of the likes of Drummond and lamented that "devout men shrink in horror from the word 'heredity,' and imagine that Henry Drummond has gone over into the camp of the enemy because he speaks of the 'Ascent of Man.'" Moreover, Hunter approvingly advertised Alfred Russel Wallace's conception of human evolution in order to underwrite a kind of doxological Darwinism. Wallace allowed natural selection to explain the emergence of the human body but found it insufficient to account for human "moral and mental faculties." Hunter's firm opinion was that readers should not "be alarmed if we see in the lower animals manifestations of some of the higher emotions which were once supposed to belong exclusively to man, as we see in man some of the selfish passions and fierce appetites of the lower animals." Plainly, Hunter was troubled neither by evolutionary gradualism, nor species transformism, nor the emergence of humans from prehominid ancestry, believing that theological resources were available to adjust to any needed conceptual realignments. Indeed, he considered that disputes about the emergence of consciousness from the somatic and the organic from the inorganic had—to all intents and purposes—already crystalized in debates between traducianists and creationists over the way the "human soul was naturally propagated along with the body by generation, instead of having been specially created (as the creationist asserts) at the origin of each new individual."[101]

Just what was it, then, that encouraged the Knox Presbyterians to seek some rapprochement with Darwinian biology when many in the Toronto scientific coterie were indifferent to it, or indeed uneasy about it? In large measure, I think, their interest stemmed from their engagement with Scottish Common Sense philosophy and their use of Baconian induction to underwrite the idea of progress rather than to outlaw speculation.[102] We have just noted Dewar's openness to working hypotheses as appropriate scientific procedure. And in prosecuting his own vision, the Knox College professor of apologetics and Old Testament Robert Yuile Thomson threw off Baconian shackles to defend the value of a priori theorizing; as he put it, "Kepler could never have discovered that the heavenly bodies move in elliptical orbits, had the idea of an ellipse not been already in his mind."[103] According to Michael Gauvreau, the rejection of Darwinism—from theological as well as scientific perspectives—was frequently rooted in the Baconian ideal.[104] The extent to which the Toronto theologians could free themselves from its tightest clutches and rework its principles in a new setting fostered fresh dealings with Darwin and his conjectures. At the same time, in their metaphysics, the Knox College fraternity set themselves steadfastly against the philosophical idealism that was becoming fashionable in certain centers of learning in Scotland. James Stalker, for instance, resisted any inclination to place a chasm between the world of perception and an external realm.[105] Others repudiated efforts to collapse subject and object or to retreat into Hegelian-inspired pantheism. To such writers, Darwinism presented a healthy empiricism about the natural order by maintaining a crisp distinction between cause and effect. Significant, too, I suggest, was the turn to the idea of progressivism that guided textual critics examining the development of the biblical documents. These maneuvers opened up what might be called a "trading zone"—to use Peter Galison's term[106]—between science and religion through which intellectual traffic could be channeled. Of course, the tradition of biblical criticism long predated Darwin's intervention, and, at least in Germany, that scholarly community registered no concerns over Darwinism when it began to circulate.[107] In Canada, it seems, evolutionary thought-forms delivered a vocabulary that facilitated the pursuit of the idea of textual evolution in some quarters and fostered the application of the notion of historical development in various arenas.

Thomson's 1890 article "The Evolution in the Manifestation of the Super-natural" shows something of how the idea of evolutionary progressivism could be readily mobilized for theological purposes. As he conceived it, the principle of evolution, which had initially developed in the study of biology, had been "successively extended to other subjects, until now it claims to embrace the phenomena and genesis of the entire universe." This was not to be lamented; it was to be embraced. For evolution had established that "everything commences in a rudimentary condition, and passes through a series of states, each slightly varied from and slightly more developed than the preceding, until a precise and determinate form is reached."[108] Such a schema suited his textual needs perfectly. His article in fact was the text of the inaugural lecture he delivered on his appointment to the position of professor of apologetics and Old Testament the year before, and it gave voice to his own sympathetic stance on the new biblical criticism. For that task, evolution proved to be extremely valuable.[109] Calling on the support of Henry Drummond, Thomson advanced a kind of evolutionary metaphysics suggestive of the vision later promulgated by Teilhard de Chardin. As he understood it, nature displayed "an orderly progress and increasing development," namely, "an evolution" that incorporated ever greater manifestations of the supernatural.[110] The idea had enormous potential. When coupled with the concept of fields of force introduced by Clerk Maxwell, it provided Thomson with a useful vocabulary in which to conceive of the emergence of the human from the natural, the progressive revelation recorded in the Old and New Testaments, and the advent of the morality of self-sacrifice. All of these convictions resonated with the intellectual energies of the leadership of the Scottish Free Church for, as Fraser observes, the "mental, moral, and spiritual evolutionism of Thomson would have been reinforced in his biblical studies at New College, Edinburgh, with A.B. Davidson and Marcus Dods."[111]

Something of this sentiment also came to the fore in an 1895 piece, "Evolution of Scripture," by Rev. John Thompson. Critical of the Koran as a document "made up by mechanical additions with no progress"—compared with the biblical record, which unfolded what he called a "scheme of advancing doctrine"—Thompson portrayed the Hebrew scriptures as "a wonderful study in evolution." To him, the biblical text "evolved and grew in connection with the history of our race." Drawing sustenance, like Robert Thomson, from analogies with the natural world, he conceived of revelation as a "gradual unfolding of a Divine purpose; a progress of doctrine, from a central principle, all coming into clearer and fuller manifestations." This maneuver enabled him to face head on the moral dilemmas that confronted readers of the Old Testament, with its genocide, tribalism, and obscure rituals; to Thompson its record could now simply be read as embodying the "imperfect and elementary conceptions of God and of morality" that characterized an earlier phase in the evolution of religious culture.[112] Here the idea of evolutionary transformism open up a zone of exchange with textual history that could be exploited. For his part, Alexander Blair found value in Herbert Spencer's Unknowable Force and urged readers to approach the idea "with a little less prejudice than we are accustomed to." If knee-jerk narrow-mindedness could be avoided, he reckoned, thoughtful church members might come to realize that "the divergence between the Inscrutable Power of the scientist and the God of the Christian might not be very wide or dangerous."[113]

In these ways, the Knox College fraternity worked hard to conjoin orthodox theology with Darwinian science—what the Scottish-born principal of the college, William Caven, depicted as "clerical conservatism and scientific radicalism." In charge of Knox since 1870, Caven did all in his power to marshal both higher criticism and evolutionary theory in the cause of a moderate evangelical creed; church members and scientists alike should be as unwilling victims of "an unreasonably timid conservatism" as of "presumptuous radicalism."[114] Whereas many of Canada's natural scientists mobilized their Baconian heritage to resist, on inductive grounds, the idea of evolutionary transformism, Caven engaged in a productive dialogue with the Baconian imperative to historicize theology and underscore the principle of evolutionary progress. Operating within a broadly Baconian framework, he nonetheless lamented the inclination of those too inclined to judge "moral evidence" at the bar of the canons of induction. To him, born-again Baconianism, to use Zeller's phrasing,[115] did not proscribe evolutionary development; it facilitated it. At the 1877 Knox College commencement, Caven insisted that theology must be "a progressive science" and that, while in its most fundamental essentials it remained constant, it was nonetheless "continually changing by the growth of new truths."[116] Simple empirical history showed that much. Being inductive about the development of theology itself opened the doors to a progressivist mindset. In this way, as Gauvreau observes, Caven "adapted 'historical theology' to accommodate the powerful Baconian imperatives," though he did so in a way that promoted, rather than prohibited, the adoption of the principle of evolutionary change.[117]

This enlistment of evolutionary rhetoric on the part of the intellectual leadership of Toronto Presbyterianism at Knox College displays a notable willingness to engage in a creative conversation with evolutionary theory. Not only were the particulars of Darwin's theory of evolution—including human evolution—treated with equanimity, but the idea of evolutionary progress was melded with a longer-standing tradition of developmentalism to serve as a potent rhetorical resource to historicize revelation, religious experience, and theological conceptualization. These projects render deeply problematic the assumption that the anti-Darwinian missiles launched from John William Dawson in Montreal were typical of Canadian Presbyterian responses to the new evolutionary biology. If Dawson was, to some degree, out of line with the measured, if critical, judgments on Darwin issued by most of his Toronto scientific colleagues, his resolute anti-Darwinism was much more profoundly at variance with the stance adopted by his fellow Scots Presbyterians at Knox College.

◆ ◆ ◆

IF CARL BERGER'S ANALYSIS—an account that dwells on the remarkable absence of engagements with Darwin among Canadian natural historians more generally— is well founded, then there may be good reason to suppose that in Toronto evolutionary motifs fared rather better in certain theological circles than in scientific ones.[118] In large measure, both of these readings of Darwin's proposals—whether supportive or suspicious—were molded in the crucible of a Scottish-shaped Baconian legacy. Among natural scientists, indeed, the entire era between the Rebellion of 1837 and 1920 has been characterized as "The Scottish Practical Science Period." As Jerry Melbye and Chris Meiklejohn acknowledge, scientific endeavor during these decades was oriented toward "practical solutions to better exploit the soil or the environment."[119] Such preoccupations made scientific practitioners, to a large degree at any rate, indifferent if not hostile to theoretical speculation, to proposals that moved any distance from inductive particulars, and to the elaboration of theoretical edifices. Science was about the dogged digging for data, the subservience of scientific endeavor to the values of practical utility, and the patient expansion of knowledge by small-scale incremental advance. These sentiments found forceful expression in the response of the Montreal-based geologist John William Dawson. In Toronto, they also governed the more muted reaction to Darwin's proposals. As time went by, Darwin secured followers there, too, but the measured rejection of some natural scientists and the even more conspicuous silence of others attest to the continuing grip that Bacon's legacy exerted on them.

Among the Knox College fraternity, the use of induction for theological purposes reinforced in that community a sense of historical consciousness already well developed through their acquaintance with the new textual methods and source criticism circulating among the intellectual leaders of the Scottish Free Kirk. Central to these concerns were ideas about progressive revelation, the evolution of religious ritual, the growth of spiritual consciousness, and the developmental nature of textual composition. All of these resonated with the transformism of

Darwin's account and produced a more creative rendezvous with the idea of the evolution of species and the emergence of the human race from earlier organic forms. Evolution provided the Knox College circle with a rich source of metaphor in which to couch their theological insights. In contrast to their coreligionists in Belfast, Toronto Presbyterians embraced, albeit in a teleological form, evolutionary rhetoric for both scientific and theological ends. Darwinian evolution suited their domestic agenda, and in using it for their own purposes, they steered their own course between those who repudiated it for religious reasons and those who championed it for naturalistic ends.

CHAPTER FIVE

Columbia, Woodrow, and the Legacy of the Lost Cause

W OODROW WILSON'S uncle, James Woodrow (1827–1907), was a staunch Presbyterian, a firm believer in the "divine inspiration of every word" in the Bible, and a self-proclaimed advocate of its "absolute inerrancy."[1] Ironically, it was also his fate to find himself immortalized on the side of the scientific angels over against the benighted forces of religion in Andrew Dickson White's famous *History of the Warfare of Science with Theology in Christendom.* What secured his canonization in White's gallery of scientific martyrdom at the hands of religious obscurantism was his dismissal in 1886 from the professorship he had held for over a quarter of a century at the Southern Presbyterian Theological Seminary in Columbia, South Carolina, on account of his views on Darwin's theory of evolution. Exulting in the "self-condemnatory" declarations of Woodrow's assailants, White echoed the sentiments of a "thoughtful divine of the Southern Presbyterian Church" who judged the actions of the church authorities to be "vicious and suicidal." Their tactics amounted to using "dynamite . . . to put out a supposed fire in the upper stories of our house"—with "all the family" still inside![2] In the June 1894 issue of *Popular Science Monthly,* White had already staged the Woodrow affair as a "striking" illustration of how the *Origin of Species* had thundered into "the theological world like a plow into an ant-hill" and how its guardians "who were thus rudely awakened from their old comfort and repose had swarmed forth angry and confused."[3]

The whole affair has been described as "the greatest controversy the Presbyterian Church, U.S., has ever known."[4] It was a grand public spectacle, which, at the time, made headline news. The *New York Times,* under the banner "Woodrow's Heresy Trial," reported that the "largest congregation so far of the General Assembly" gathered for the hearing.[5] The southern correspondent for the *New York Observer* told his readers that Columbia had no use for "tadpole theology" of the type that Woodrow was broadcasting—an evident reference to human evolution from primitive life forms.[6] As Dr William Adams, one of the leaders for the prosecution in its later stages, announced during the heated exchanges: "Then another long period intervenes, and they get a tad-pole . . . and this little fellow somewhere, somehow gets ashore, and that, Mr Moderator and members of the assembly, was the landing of your ancestor."[7] To the *Springfield Daily Republican*, which reported the story, "Mr Adams' humor was the humor of the insolent and his fear of science the fear of the ignorant." Not everyone agreed. Whereas this correspondent was sure that it was only "unread and ill-informed exhorters" who saw in "the discovery of a grand and beautiful law in the natural world any menace to pure morals or religion,"[8] the General Assembly that convened in Columbia in 1888 detected in evolution an infidel canker that would rot the entire fabric of southern culture. By contrast, to Woodrow's yet-to-be-famous presidential nephew, Woodrow Wilson, the whole episode bordered on the burlesque. Writing from Johns Hopkins, where he was carrying out graduate work, Wilson told his fiancé Ellen Axson in June 1884: "If Uncle J. is to be read out of the Seminary, Dr. McCosh ought to be driven out of the church, and all private members like myself ought to withdraw without waiting for the expulsion which should follow belief in evolution. If the brethren of the Mississippi Valley have so precarious a hold upon their faith in God that they are afraid to have their sons hear aught of modern scientific belief, by all means let them drive Dr. Woodrow to the wall."[9]

Figuring out just what was at stake in this drama with its serpentine twists and turns is our quarry in this chapter. These machinations expose how Darwinism was constructed in and around the dispute over evolution at the Columbia Seminary during the final decades of the nineteenth century and what was believed to be in jeopardy if evolution's empire were to expand in Dixie.

Tadpole Theology on Trial

The immediate cause of Woodrow's undoing was the stance he had adopted in a lecture he delivered in 1884 to the Alumni Association of the Columbia Seminary. Suspicions about Woodrow's attitude to Darwinism had been circulating for several years, and a new member of the college's board of directors, Rev. Dr. Joseph Bingham Mack (1838–1912), reporting the rumors he himself had heard—"irresponsible whisperings" as one commentator put it[10]—urged Woodrow in 1879 to clear the air by putting his views on the whole subject into print.[11] Woodrow declined. He was "too busy to notice the slanders" of troublemakers like "old Plumer," he quipped, evidently referring to his former colleague Dr. William S. Plumer, with whom he'd had a running spat for quite a few years.[12] That didn't satisfy Mack. He took Woodrow's brush-off as a sign of caginess and got to work on the other members of the seminary's board of directors to adopt a resolution that would force Woodrow "to give fully his views as taught in this institution upon Evolution, as it respects the world, the lower animals and man."[13]

The gauntlet had been thrown down, and the spring meeting of the seminary's alumni association was set as the venue for Woodrow's statement. On 7 May 1884, a Wednesday, Woodrow delivered his judgments in an address simply entitled "Evolution."

Woodrow was well equipped to declare on evolution's scientific credentials. The son of a Presbyterian minister who served in Canada and Ohio, he had come as a child to North America from his birthplace in Carlisle, England, and in 1853 secured an appointment as professor of natural sciences at Oglethorpe University, a small Presbyterian school in Georgia. During his time there he studied for a summer with Louis Agassiz at Harvard and later spent a couple of years at the University of Heidelberg, which awarded him a PhD in 1856.[14] While teaching at Oglethorpe, he was ordained to the Presbyterian ministry in 1859, two years prior to being offered the new Perkins Professorship of Natural Sciences in Connection with Revelation at the Columbia Seminary.

Woodrow began his discourse by making clear that he saw his present task as reflecting on the theological implications of evolution, rather than delivering a dissertation on scientific data pertaining to the subject, and he outlined his own thinking on the relationship more generally between the teachings of scripture and the findings of science. The burden of these introductory remarks was to reinforce John Calvin's dictum that the Bible "does not speak with philosophical acuteness on occult mysteries . . . He who would learn astronomy, and other recondite arts, let him go elsewhere." The upshot was crystal clear: "The Bible does not teach science: and to take its language in a scientific sense is grossly to pervert its meaning." But this did not undermine its authority. Woodrow was confident that he could affirm the plenary inspiration of scripture, down to "every word," even while adopting the theory of evolution.[15]

Woodrow's tactic was to insist that evolution simply described the derivation of organic beings from previous life-forms without "any reference to the power by which the origination is effected; it refers to the mode, and to the mode alone." This definitional move, he believed, exposed the incoherence of asking "whether the doctrine is theistic or atheistic, whether it is religious or irreligious, moral or immoral." Posing such questions was simply a category mistake. "It would be as plainly absurd," he went on, "to ask these questions as to inquire whether the doctrine is white or black, square or round, light or heavy. . . . These are qualities which do not belong to such subjects." Woodrow's strategy, in essence, was to resolve conflict by boundary maintenance: science dealt with one territory, religion with another. As for specific scriptural passages that seemed to militate against an evolutionary reading of natural history, Woodrow worked hard to liberate a variety of texts from the presumptions that underlay traditional readings. On the creation of Adam from the "dust of the ground," to take just one example, Woodrow examined how these words were used elsewhere in the biblical narrative to demonstrate their metaphorical possibilities.[16] His aim was to defend the proposal that the body of Adam may have been produced by evolutionary processes from pre-human life forms, though he insisted repeatedly—in a move originally advocated by the Catholic naturalist St. George Mivart in his 1871 *The Genesis of Species*—that the soul was the product of immediate divine creation.[17]

Woodrow's case for the defense, however, did not stop with simply showing that evolution could be made compatible with theology. He went much further than that by laying out evidence in support of the theory, despite affecting not to dwell on scientific data. Skeletal homologies like the architectural similarities between human and animal skulls, for example, were entirely understandable on the assumption of common descent. The existence of rudimentary organs was precisely what would be expected if the doctrine of descent with modification were true. The morphological transformations of the human embryo, which bore a striking resemblance to evolutionary progress "from the lowest to the highest," were "just such as the doctrine of descent requires." And the geographical distribution of living beings was likewise what an evolutionary history would be likely to deliver: "if the theory of descent with modification is true, it should be expected that in the regions recently separated, the animals would differ but slightly; in regions separated long ago, the animals would differ more widely; and that, just in proportion to the length of separation. This is exactly what we find." To any fair observer, Woodrow was convinced, such considerations should "almost compel belief in the doctrine." Because to him evolution was another term for "mediate creation"—the view that God's creative activity in the world was mediated through secondary agencies—he could not see "how any one could hesitate to prefer the hypothesis of mediate creation to the hypothesis of immediate creation." As for the objection that evolution degraded the human race by tracing it to animal forebears, Woodrow was far from convinced that "dirt is nobler than the highest organization which God had up to that time created on the earth."[18]

Over the following summer months, Woodrow's address seriously stirred up the hearts and minds of readers of the religious press. In June, the *Christian Observer* carried critical editorials noting that Woodrow seemed to have changed his mind over the previous decade and that a seminary teacher had no right to do so without informing his public of his new stance.[19] Opposition to Woodrow was also prominent in such denominational magazines as the *Central Presbyterian*, the *North Carolina Presbyterian*, the *Southwestern Presbyterian*, and the *Texas*

Presbyterian. The latter insisted, for example, that on the face of it, it was hard to believe that when Genesis declared, "The Lord formed man of the dust of the ground," it really meant to say, "Man was born of an ape by ordinary generation."[20] Antagonistic correspondence continued throughout August, September, and October with a trickle of rebuttals from contributors to the *Southern Presbyterian Review*, which was under the editorial control of Woodrow himself.[21] James L. Martin, formerly a medical doctor, for example, was sure that the seminary needed men like Woodrow who were both scientifically literate and theologically competent. But voices of support were few and far between as the language of hostility hardened: "heresy," "mischievous," "godlessness," "sensuality," "minds already degraded," and the like began to litter the printed page.[22] It became increasingly clear that there were growing worries over the effect Woodrow's presence would have on the seminary's student enrollment. One writer flatly stated that no presbytery would send ministerial candidates to an institution promoting evolution.[23] This stance was certainly in the spirit of Mack's long-standing irritation with Woodrow. Shortly after hearing the address, Mack, who had oversight of the seminary's finances, wrote to a friend: "The brethren in the Mississippi Valley feel that they cannot and will not support Columbia Seminary if Dr. Woodrow remains there as a professor . . . they cannot trust their young men with us . . . he is a burden too heavy for us to carry; I will not hesitate to say that the issue is between the welfare of the Seminary and the retention of Dr. Woodrow."[24]

Although Woodrow's enemies succeeded in whipping up widespread antagonism toward him, even his most vociferous opponents doubted that a charge of heresy could be made to stick. At the September meeting of the seminary's board, Rev. James Stacey tabled a resolution condemning the idea that Adam was descended in any way from lower animals and forbidding the Perkins Professor to teach any such theory. It was voted down by a margin of eight to three. Instead, a statement welcoming Woodrow's stance on scripture was passed.[25] The committee dealing with the seminary brought in majority and minority reports, the former agreeing that the theory of evolution was a scientific matter and that the church had no obligation to declare on its truth or falsity. Among its resolutions was the judgment that "the relations subsisting between the teachings of Scripture and the teachings of natural science are plainly, correctly, and satisfactorily set forth in said address." Though it went on to note that it was "not prepared to concur in the view expressed by Dr. Woodrow as to the probable method of the creation of Adam's body," the board declared that "there is nothing in the doctrine of Evolution, as defined and limited by him, which appears inconsistent with perfect soundness of faith."[26] The minority report, by contrast, insisted that

Woodrow's views were "contrary to the interpretation of the Scriptures by our Church and to her prevailing and recognized views." The nub of the problem for these dissidents was that Woodrow's views unsettled "the received interpretation of many passages of Scripture."[27] Regardless of the status of Woodrow's standpoint when judged against Christian theology *simpliciter*, the accusation that he was violating the denomination's local traditions, as we will see, was far from insignificant. Still, for the meantime, Woodrow had escaped. But he was soon to learn to his bitter cost that winning a battle was not the same as winning the war. The campaign had just begun.

The denomination did not generally receive the board's actions with enthusiasm. As Thompson observes, "The *Texas Presbyterian* was 'surprised,' the *Central Presbyterian* 'amazed,' the *North Carolina Presbyterian* 'saddened' by the support given to Dr Woodrow." Patience was fraying and tempers wearing thin. According to one minister, the "Perkins Professor has put the dim figure of an ape on the bunting that waves above his chair . . . Will the General Assembly also put the *mark of the beast* on the blue banner of Presbyterianism?" Another correspondent, a woman, posed the question: "Did Christ foreordain to offer himself for an animal in process of evolution?"[28]

During the months that followed, the tangled networks of southern ecclesiology were buzzing with the news. The presbyteries and synods of Georgia, North Carolina, Alabama, Virginia, Kentucky, Nashville, and many more were determined to have their say. Most significantly, in October 1884, the South Carolina synod—which received the two reports of the Committee on the Columbia Seminary—invited Woodrow to speak in his own defense. Woodrow's two-day presentation seems to have been entirely unscripted: the whole debate apparently lasted for five days, with Woodrow speaking for a solid five hours one evening![29] It was a grueling business. During his speech, having survived several proposals to adjourn proceedings, he was eventually forced to concede, "I am so exhausted that it is utterly impossible for me to proceed." He had to postpone the remainder of his address until the following evening. The published version, which extended to over sixty printed pages, was stitched together from stenographers' reports and apparently omitted "much that was said by the speakers, as well as many questions put by members of the Synod and short speeches made by them, while Professor Woodrow occupied the floor."[30]

Woodrow began by reviewing the history of the whole affair, digging back into the early days of the Perkins Chair, and spending a good deal of time niggling over just what he was on trial for, if indeed a trial it really was. He worked particularly hard to exploit the ambivalence in the entire proceedings by affecting puzzlement over just what he was charged with, since even the minority report flatly stated that "the question whether Dr Woodrow's views in regard to evolution involve heresy is not before the Synod." He was also at pains to remind his hearers that when he was originally appointed to his chair a quarter of a century earlier, the board knew perfectly well that he "was going to teach something very different from the doctrine that the world was created only one hundred and forty-four hours before Adam."[31] He did not miss the chance, moreover, of capitalizing on the ill-judged suggestion of Dr. W. F. Junkin that the actions of Catholic officialdom over Galileo might profitably be imitated in Woodrow's case. Woodrow seized on the opportunity this gave him to read a lengthy extract from the original sentence passed on Galileo more than 250 years earlier. The passage spoke for itself, but at least, Woodrow mused, Galileo had been extended the privilege of being put on trial according to the legal constitution of his church.

Woodrow's lengthy statement was woven through with several salient themes. First and foremost, his defense rested on the fundamental importance of distinguishing between matters of science and matters of faith. To Woodrow, a good deal of trouble about the whole subject sprang from a failure to recognize that good fences make good neighbors. Science dealt with one sphere of reality, theology with another. He reiterated the fundamental importance of boundary maintenance. What was at stake was nothing less than the freedom of science. The Bible, to put it bluntly, did "not teach science." To expect professors to inform a college board that they had changed their minds and their curricula in the light of growing evidence on some scientific subject was nothing short of "preposterous." Ecclesiologically, this meant that church leaders had no business adjudicating on anything other than issues of faith. His accusers' proclamation that evolution was an "unverified hypothesis" was thus an act of territorial transgression. In their capacity as church representatives—whatever might be their private opinion—ecclesiastical courts, clerical judges, college trustees, or indeed assembly moderators had no right to "utter any sound on that subject." On scientific questions it was the church's duty to rely on the testimony of experts. But to enjoy the benefits of their expertise, they had to be left to pursue their inquiries untrammeled. To do anything else was like telling lawyers how to behave in court or physicians what their diagnoses should be. Clergymen, Woodrow contended, were the "professional counsel" of a church in matters of scriptural teaching, but that was the limit of their proficiency.[32] Beyond lay regions under the authority of other competent professionals.

Second, it plainly bothered Woodrow that his adversaries focused their complaint on his alleged violation of "the interpretation of the Scriptures by our Church and . . . her prevailing and recognized views." Woodrow eagerly fastened on the wedge that this formulation drove between local and universal Christianity. To accept his accusers' charge was to acknowledge "that the highest and absolute sense of the Sacred Scriptures is different from that which you pledge yourselves to teach as ministers." "I am under no more obligation," he went on addressing the moderator, "to teach received interpretations than you are."[33]

Along the way, Woodrow gestured toward many other subjects. He reiterated his belief in scriptural inerrancy. He affirmed that the unity of the human race was not in any way jeopardized by adopting an evolutionary account of Adam's formation. Indeed, on this subject he did not fail to miss the point that it was the antievolutionary Louis Agassiz, so beloved by many for his creationist stance, who truly compromised humankind's unity by his assertion that the different human races had descended from different original stocks. At the same time, he marshaled a host of natural scientists—Alexander Agassiz, Edward Drinker Cope, James Dana, Asa Gray, Othniel C. Marsh, Alpheus Packard, Addison Verrill, and many more in the United States, never mind Europe—as allies in support of his pro-evolution stance. He also insisted that because he accepted evolution that did not mean he endorsed everything that Darwin or Spencer or Haeckel had to say. As for any attempt to resolve theological questions by head count, Woodrow found such proposals absurd: in the American South, he wryly pointed out, Baptists would outvote Presbyterians on any matter they chose by a factor of ten to one! And he took full advantage of the opportunity the hearing afforded him to point out that his assailant, John Lafayette Girardeau (1825–1898), despite their long association as seminary colleagues, had never broached the subject with him in private conversation.

It was at the Greenville synod, in fact, that Girardeau emerged as Woodrow's chief antagonist as he took on the mantle of chief prosecutor. Since 1876 he had occupied the chair of didactic and polemic theology at the Columbia Seminary and was widely known as a doughty defender of orthodoxy against the inroads of modernism in all its shapes and forms.[34] From the outset, Girardeau, in two speeches lasting some three hours, made it clear that the issue before the denomination was not technically a charge of heresy—Woodrow, after all, had not renounced any fundamental doctrine. In a critical sense, the whole matter for Girardeau reduced to a single issue: "The question which, in my judgment, is really before the Synod is in regard to the relation between Dr. Woodrow's hypothesis and the Bible as our church interprets it: between the scientific view and our Bible—the Bible as it is to us." Before getting to that though, he entertained his hearers at some considerable length on the apparently fundamental difference between "non-contradiction" and "the harmony of non-contradiction" in matters of science and theology. As he saw it, this distinction was vital to understanding the precise remit of the Perkins Chair, and it led him into a fairly tortuous exposition of the law of the excluded middle, how theology should deal with "a proved truth of science" compared with "unverified scientific hypotheses," and such like.[35]

But it was Woodrow's departure from Southern Presbyterian convention that he returned to again and again as the lynchpin of his case for the prosecution. In large part Girardeau's disquiet centered on what he saw as a church's right to determine precisely what its own seminary should teach. The question was not about orthodoxy in any absolute sense; it was simply about the need to ban the teaching of anything "contrary to our church's interpretation of the Bible." To him what was at stake was the heritage of an entire culture. When a church permitted a seminary teacher to promulgate beliefs out of step with its own tradition, he ominously warned, "she actually makes arrangements for the overthrow of her own views. She arranges for her own sacrifice." For any church to "surrender" to "unverified hypotheses" was an altogether "wretched" business. But even "a proved truth of science," Girardeau declared, "ought not to be inculcated in a theological seminary when it contradicts our Standards." The Woodrow case was plainly about a lot more than evolution; it was about the very survival of the entire edifice of Southern Presbyterian culture. Any professor failing to accede to the strictures of traditionalism was guilty of intolerable rebelliousness. What was the much-vaunted principle of freedom of expression, after all, "but insubordination to law in high places, and the encouragement of the temper of insubordination to law in those who are to be its expounders and defenders"? If the denomination allowed Woodrow to continue to teach evolution, it would be complicit in its own subversion. Besides all this, the church footed the seminary's salary bill, and he who pays piper calls the tune: "The church," Girardeau slipped into a footnote, "cannot be expected to pay for teaching to which she is conscientiously opposed."[36]

Girardeau now took to applauding the wisdom of his denomination's traditional conservatism in resisting a range of scientific innovations. It is highly significant that his eye first fell on the question of the "Specific Diversity of the Human Races as opposed to the church's doctrine of the Unity of the Race." That beastly doctrine had lately been touted with just about as much gusto as the current evolution hypothesis, he noted. Mercifully, the church had held its ground for, as we shall later see, acquiescing to this anthropological novelty would have been a profound blow to how the denomination negotiated the question of race relations in both the ante- and postbellum South. It was something similar for the challenge of the "extreme Antiquity of Man," which, he recalled, was "at variance with the church's view of the biblical chronology." Spontaneous generation and original linguistic plurality were every bit as wisely resisted. But he went further yet. His catalogue of outlawed topics that contravened "the requirements of conservatism and consistency" displayed a markedly greater adherence to biblical literalism than any comparable Scottish Presbyterian culture. Girardeau's syllabus of errors, for example, included any teaching that contradicted the standard understanding of the Genesis narrative as proclaiming that the world had come into being through a set of discrete creative acts over the space of six days.[37]

All of these were pernicious assaults on the South's settled certainties but none more so than Woodrow's troubling the traditional interpretation of Adam's creation. In every way imaginable, Woodrow, by allowing for an evolutionary account of his corporeal formation, had upended the dogma that "Adam's body was formed of dust by a sudden, supernatural, constructive act of God." The disgust, Girardeau assured his listeners, was universally felt. The "overwhelming mass" of southern Presbyterians was "opposed to the hypothesis of the Perkins Professor." The situation couldn't be more grave. But just in case anyone should remain in doubt, Girardeau ended with alarmist, not to say inflammatory, rhetoric intended to further heighten tensions: "I have never believed heretofore that the foundations of the Seminary were seriously endangered. Even in its darkest days I trusted that the kind Providence which had favored it from its beginning would continue to sustain it. But now I feel that the institution is on the edge of deadly peril . . . Let the hypothesis of evolution be inculcated in the theological school at Columbia . . . and the majority of the people of God will withdraw from it their sympathy and their support."[38]

Voices in support of Woodrow were quickly forthcoming. John B. Adger, former Armenian missionary and seminary professor, was of the opinion that the "genius" of any seminary was precisely that it engaged in "inquiry into all truth" and not simply some set of predetermined convictions. C. R. Hemphill, a current faculty member, expressed concerns about the implications of a policy such as Girardeau's since it left the seminary in the embarrassing position of describing a theory that "every scientific man believes" as only "an 'unverified hypothesis.'" W. A. Clark drew attention to the Princeton theologian A. A. Hodge's observation that evolution "could only be regarded with the most friendly interest." And James L. Martin, the minister and medical man we have already encountered, insisted that Woodrow equipped his students to resist the "assaults of scientific infidelity" and recollected how he himself had been prevented by "the lectures in

Dr. Woodrow's classroom" from spiraling "in a wild downward career to infidelity and atheism and cheerless blank despair."[39]

Regardless of these supporting pleas, when it came to voting, the synod chose to discard both the majority and minority reports and resolved by a margin of fifty to forty-five to outlaw the "teaching of evolution in the Theological Seminary at Columbia, except in a purely expository manner, with no intention of inculcating its truth." Woodrow's supporters claimed that was all he had ever done, but it was a Pyrrhic victory at best. In a telegram that found its way into the hands of G. B. Strickler, Girardeau put it straight: the South Carolina judgment was no compromise at all; it was "distinctly and intentionally anti-Woodrow."[40] Strickler used Girardeau's message to maximum effect. At the synod of Georgia a few weeks later, he quoted from it in a speech that described evolution as "horrible to thousands of people" and goaded his hearers with the taunt that members supporting evolution might as well put an ape on "the banner" of their faith and "write the obituary of the church." He had caught the mood. W. E. Boggs, a Woodrow supporter, tried to call on the august authority of the Edinburgh theologian Robert Flint, "the ablest man in Scotland," as a compelling witness for the defense. But the tide was running fast in the opposite direction. Strickler ominously warned that influential congregations were now threatening to "not give a cent" toward the work of the seminary unless actions were taken. His motion that "the teaching of evolution as contained in Dr. Woodrow's address be disapproved" and that the board be directed "to take whatever steps may be necessary to prevent it" was adopted with overwhelming support—sixty to twenty.[41] In other synods, too, opposition was now rising. In South Georgia and Florida, and in Alabama, the vote went decisively against Woodrow.[42]

The writing was now on the wall. When the substantially new board of directors next met in December 1884, it called for Woodrow's resignation. He declined the invitation, and the board voted to remove him from his chair. The judicial maneuverings that surrounded Woodrow's subsequent appeal, on the grounds that the board had no authority to evict him without due process from his chair, need not detain us. Suffice to note that, as Thompson puts it, "the debate over the evolutionary hypothesis continued, becoming more heated with each passing week."[43]

By now, other prominent protagonists had entered the listings. For example, George D. Armstrong (1813–1899), formerly a professor of chemistry and geology at Washington and Lee University, currently a Presbyterian minister in Norfolk, Virginia, and a controversialist of some reputation, had presented his thoughts on the subject of evolution in 1884. He would soon emerge as a key opponent of Woodrow during the Augusta General Assembly and increasingly became the denomination's spokesman on matters of science and religion. Here he concluded that "The Evolution hypothesis when taken in its widest range . . . is, beyond all question, atheistic . . . In this form evolution is confessedly irreconcilable with the Bible and our Christian Faith." In particular, Armstrong was certain that "its account of the origin of man . . . is irreconcilable with . . . Bible declarations."[44]

On the other side of the divide stood John William Flinn (1847–1907), brother-in-law of John Adger (another Woodrow supporter). Flinn was a minister in New Orleans and later a rather unsuccessful professor of moral philosophy at South Carolina College. In an article for the April issue of Woodrow's *Southern Presbyterian Review,* he lamented how, in an age "morbidly alive to the apparent collision and antitheses of science and religion," his own denomination had done a "cruel wrong" to James Woodrow. A "spirit of unreasonable jealousy and fear towards scientific inquiry" had dogged the whole affair, and "rash and harsh judgments" had been summarily dispensed by those who should have known better. All that Woodrow's foes had managed to achieve was to place "another javelin . . . in the hands of future John W. Drapers."[45] A profound misunderstanding of the nature of hypotheses, Flinn urged, had afflicted the Woodrow hearings, and he labored long and hard, quoting from authorities like Aristotle, Isaac Newton, William Hamilton, and John Stuart Mill to substantiate their indispensability to scientific progress. In a second piece, which appeared in the July edition, he sought to show how his denomination was staggeringly out of step with the overwhelming mass of scientific opinion. In an extensive catalogue of individuals and institutions favoring evolution, Flinn amassed the supporting testimony of many of those scientists that Woodrow himself admired, such as Othniel C. Marsh, James Dana, and Addison Verrill at Yale; Asa Gray, Nathaniel Southgate Shaler and Alexander Agassiz at Harvard; George Macloskie and Cyril Fogg Brackett at Princeton; and many more. These supporters accompanied a wide-ranging inventory of British and European scientific advocates. All this served to place the anti-Woodrow faction of the denomination in an "awful dilemma." With "wild excitement and unreasoning prejudice," he announced, this "majority" party had put its own church in a position that could only be described as "absurdly ridiculous, were it not so pitifully deplorable."[46]

Things dragged on with appeal, counterappeal, and complex crosscurrents of one sort or another. And the whole affair began to spiral in other nasty directions, too. Woodrow's opponents mentioned his savage temper, his intransigence, his apparently poor record of chapel attendance, and the suspicion that he was alto-

gether too busy making money through publishing ventures and various directorships.[47] A formal trial was set for the meeting of the General Assembly in May
1886. The *Springfield Republican* portrayed it as "the heresy drama of the year."[48]
Girardeau, who had earlier declared that there were no grounds for accusing
Woodrow of heresy, now happily signed the charges against him. By an overwhelming 137 to 13 the report of the antievolution faction was approved, and the
recommendation that Woodrow be removed from his post was subsequently carried. This injunction, ratified during the following winter by a range of synods,
came to full fruition in December when the seminary's board formally and unan-
imously dismissed Woodrow from his chair. Further complaints were lodged, not
least about the legal probity of the proceedings, but Woodrow never returned to
the seminary to teach. Bizarrely, he was made moderator of the Augusta Presbytery in 1888, and in the years that followed he secured the post of president of
South Carolina College, having in the interim held a professorship there. Reflecting on the whole episode, which he had witnessed and chronicled in painstaking
detail, John Adger concluded, "The hypothesis of evolution, and Dr. Woodrow
along with it" were "overwhelmingly defeated." Looking across the war-torn landscape, Adger could only see "the victorious hosts of anti-evolution marching triumphantly over the whole field."[49]

As the evolution issue continued to snake its way through the highways and
byways of Southern Presbyterian churchdom, things became at once vastly more
complicated and yet more focused. Accusations of judicial irregularity surfaced,
with innuendos about personal prejudice and barefaced animosity. Flinn, for
instance, issued a searing attack on the way in which William Adams had been
licensed to act as both appellant and prosecutor at the Georgia Synod, thereby
placing far too much power in the hands of a single individual. It was nothing
short of putting "the accused a second time on trial."[50] More withering yet was
the anonymous attack of one of the denomination's own clergymen—under the
pseudonym "Gillespie"—on the General Assembly's verdict. "Gillespie" was appalled at the church's interfering in the affairs of science. It reminded him of the
time when in 1861 the assembly had dirtied its hands meddling in the politics
of secession; that had had the effect of splitting in two the entire denomination.
Girardeau's driving a wedge between "the Bible in its highest and absolute sense"
and "the Bible, as interpreted by our church" particularly galled. That was, "Gillespie" felt, a peculiarly "popish distinction." It attested to a worrying trend toward "centralization"—an obsession to control by diktat. The machinations he had
witnessed, not least from Girardeau, only confirmed in his mind that the assembly had been treated to a "*travesty* and *caricature*" of Woodrow's views; the proceed-

ings from start to finish were a tissue of malicious "slanders," "actionable libels," and defamatory "allegations." There was no concern to give Woodrow a fair hearing; "the one sole end aimed at by a number of individuals all over the church" was to get him out of the seminary "without further delay." The policy of the assembly had certainly been effective, he mused: it amounted to "hanging the accused first, and trying him afterwards."[51]

For all the labyrinthine goings-on, Southern Presbyterian nervousness about evolution congregated largely around one issue: the origin and unity of the human species. George Armstrong's report, which was adopted by the 1886 Augusta Assembly, declared the church's teaching to be that "Adam and Eve were created, body and soul, by immediate acts of Almighty power, thereby preserving a perfect race unity" and that "any doctrine at variance therewith is a dangerous error . . . and will lead to the denial of doctrines fundamental to the faith."[52] Woodrow was perplexed. In prosecuting a complaint against the actions of the Synod of Georgia in 1888, he confessed astonishment at the allegation that the evolutionary stance he adopted "contradicts or sets aside the doctrines of the unity of the human race, of the fall of Adam, and of his federal headship." He continued:

> I confess my inability to see the grounds of this objection . . . So far as I can see, the unity of the human race depends in no way upon the material of which God formed Adam or the changes through which he had previously caused that material to pass. It seems to me to depend, so far as we are concerned with it solely upon the descent of all men from our first parents Adam and Eve. If all men are Adam's descendants, is there not a "perfect race unity"? And how is that unity involved in the question whether Adam's body was created by an immediate or by a mediate act? What more than community of origin, descent from the same pair, can be needed to "preserve the perfect race unity," which we all believe to exist?[53]

Perhaps, he speculated, what lay behind it was the "supposition" that "there must have been many men simultaneously created; for the same causes which produced one body suitable for transformation into a man would have produced great numbers of similar bodies. This is supposed to be a good logical inference from my teaching. But the fact is instead that it betrays an entire ignorance of what is involved in the doctrine of descent with modification. The first principle of that doctrine is that the modification appears in a single individual and not in many."[54]

Others were similarly puzzled. In his blistering attack on the entire business, the anonymous "Gillespie" couldn't understand how human unity was in any way affected by whether Adam and Eve were the product of mediate or immediate creation. He failed to see, moreover, "how the perfectness of the unity of our race is made in any way more certain by the deliverance of the Augusta Assembly."[55] Flinn was no less bewildered. He composed a footnote, stretching over five pages, on the subject of "Race Unity and the Causes of Race Varieties." Here he provided a pretty comprehensive synopsis of the debate over whether the human race was of monogenetic or polygenetic origin. On the plurality side of the question stood Isaac la Peyrère, Armand de Quatrefages, Louis Agassiz, Jean Jacques Rousseau, and Tom Paine. But the bulk of his survey was geared to presenting the scientific proponents of monogenism, not least those from the American South, including John Bachman, Thomas Smyth, and James L. Cabell.[56] A litany of supporting voices from naturalists, physicians, and ethnologists pressed him to the conclusion that "no scholar of any eminence . . . denies that race varieties have been produced by natural causes." All of this was marshaled in opposition to Agassiz, whose rejection of evolution was so valorized by the anti-Woodrow brigade, and to demonstrate that it was actually the *creationism* of "Pope Agassiz" that subverted "a unity of origin from one pair," not evolution, which simply demonstrated how human varieties could be derived through descent with modification from a common origin. In fact, he urged that Agassiz had recognized that if "race peculiarities . . . were produced by natural causes," that would constitute "a proof of race unity, and therefore of Evolution."[57] As we saw in chapter 1, it was precisely the perceived monogenism of Darwinism that was so troublesome to the Charleston polygenists who therefore consistently rejected it. Neither the clergy at Columbia nor the naturalists at Charleston would have any truck with Darwinism—but for entirely contrasting reasons. The floor of the General Assembly and the corridors of the Charleston Museum were very different spaces and constructed the meaning of Darwinism in very different ways.

For all that, the idea that evolution shattered human unity was commonly felt among the Southern Presbyterians and a resolutely literalist reading of the Mosaic narrative seen to be the only way to preserve intact that cherished doctrine. Why? Armstrong's 1885 declaration in his brace of lectures on evolution provides a critical clue. Central to the scientific claims of Darwinism, Armstrong noted, was organic variation and the degree to which it could be inherited in succeeding generations. Armstrong culled the literature on the subject to support his conclusion that variation occurred within definite limits and ordinarily was not passed on to succeeding generations. To be sure, controlled breeding had induced stunning changes in domestic animals, generating within a single species differences as great as the Flemish dray horse and the Shetland pony. But these were emphatically within species limits and were, in any case, subject to reversion to type.

The permanence of species was, to Armstrong, a fundamental law of nature. To substantiate it, he returned, tellingly, to the antebellum writings of John Bachman (1790–1874), Lutheran clergyman, distinguished Charleston man of science, and collaborator of John Audubon. As Armstrong recalled, it was at a critical moment in southern history in the lead-up to the Civil War that the species question had surfaced with a vengeance, since "the unity of the human race . . . was involved in the slavery question."[58]

For Bachman, the most critical arena in which the nature of species played out was encapsulated in the title of his most famous work, *The Doctrine of the Unity of the Human Race Examined on the Principles of Science* (1850). Trekking through a complex literature on hybridity, domestication, species limitation, acclimatization, sterility, albino variation, heredity, and the like, Bachman mounted his case with forensic precision against the polygenetic anthropology of Louis Agassiz, Samuel Morton, Josiah Nott, and Hamilton Smith. All pressed him to the conclusion that species were indeed permanent and that all the human races were nothing more than variations within a single species. To Bachman, racial polygenism threatened not only the authenticity of scripture but also the ideological fabric of what he considered southern Christian civilization. None of this, of course, presumed the natural equality of all race types. The son of a slaveholder and a slave owner himself, Bachman was as convinced of black inferiority as any other southerner. To him, the African was "an inferior variety of our own species," "incapable of self-government," and in need of white "protection and support." But that did not "exclude the negro from the species to which we belong," Bachman urged, and he mobilized his science to confirm the Genesis account of universal human descent from Adam and Eve. "The creation of the first human pair," he declared, which "must be admitted by all who are not atheists, was a miraculous work of God. No combination of atoms, or any general elevation of lower animals into higher orders—according to the absurd theory of La Mark . . . could ever have produced man in any other way than by miraculous power." Science was at one with revelation; both attested to the truth that God had "made of one blood all nations of men." This declaration meant that "efforts to degrade" Africans "into a different species" were "unwise and unphilosophical" and compounded "scientific error with . . . political folly."[59]

Thirty-five years later, Armstrong enthusiastically revisited Bachman's "protracted and careful study"—produced, he recalled, "at the same time as Darwin was preparing his 'Origin of Species' "—to underscore the truth that "all natural species of plants and animals are permanent." Just as Bachman had demonstrated the scientific superiority of the biblical account of human origins over the degrading polygenism of Nott and his odious company, so it could still be marshaled against the new forces of Darwinian infidelity, not least in the church itself where some—Armstrong undoubtedly had Woodrow in mind—were happily yielding to the alien idea that "the body of man is the product of evolution, the other half, along with his soul, the product of immediate creation." To toy with any such "mongrel origin" for the crown of creation was even more outrageous than "the wildest speculations of scientists."[60]

According to Desmond and Moore, settling the question of evolution was inextricably bound up with questions of racial variety throughout the length and breadth of the nineteenth century. Again and again, they show, discussion of variation among pigeons and pigs was really a surrogate for addressing the race question. Bachman knew that only too well. In a book about human unity, the weight of print dedicated to fancy pigeons might seem odd, but as Desmond and Moore put it, "Human and pigeon breeds were all of a piece: prove that these extraordinary pigeons had all come from one ancestor, and the case would be so much easier for humans . . . It remained true that one of the best places to pick up contemporary chicken lore was a human-race book like Bachman's."[61]

In the American South, of course, these questions carried an especially forceful charge, and coming to grips with that issue requires attending to the deeper prehistory of the whole Woodrow affair. A profound sense of cultural isolation and southern exceptionalism had gripped the hearts of some of Woodrow's staunchest defenders, who feared that their denominational culture was cutting itself off from current scientific and religious opinion. William Flinn, for example, painfully noted that he belonged to a decided minority in voicing confidence in Woodrow. In any head count his party would sustain a crushing defeat, and that meant, he noted with some dejection, that one in ten church members would be condemned as heretics! It was staggering to recall the extent to which "Romish persecution has found defenders in the ranks of Dr. Woodrow's opponents on the ground that the Church of Rome was carrying out its views." Tragically, the Catholic Church had not been sufficiently self-critical, never pausing to ask, "What right had she to hold views according to which, in her opinion, the burning of Giordano Bruno and the imprisonment of Galileo were duties and logical results?" We have already noted Flinn's rehearsal of the overwhelming scientific concurrence on evolution and his chagrin that the church had been placed in an embarrassing position by *"rash, blundering synodical decrees."*[62] He was no less concerned about his denomination being relegated to the theological margins. The "consensus of Christendom," he remarked, was decidedly against the anti-Woodrow majority. Representing their closest denominational neighbors, the Northern Presbyterians, James McCosh, Alexander Hodge, Charles W. Shields, and Francis Landey Patton all considered evolution compatible with scripture. So did the Dutch Reformed Church, the United Presbyterians in Scotland, the Free Church of Scotland, the Episcopal Church in both Britain and America, and L'Eglise Réformée de France. If Woodrow's opponents were correct that his teaching was out of step with the "received interpretation" of the Southern Presbyterian Church, it only showed the degree to which the denomination had blithely detached itself from the theological mainstream.[63]

To understand precisely why these sentiments and the antimodernist conservatism they disclose were so tenaciously retained in the face of increasing cultural isolation, we need to set the whole episode in its deeper historical context. To do so we must turn the clock back a decade or so to an earlier feud in which Woodrow was embroiled during the 1870s and inspect that dispute in the light of circumstances prevailing in the late 1850s, when the Perkins Chair was first established. The same refrain, sounded thirty years earlier in the Southern Presbyterian encounter with the sciences, echoed down the decades as older guardians of the culture's sacred flame struggled to retain their identity in a fast-changing social situation. The uncompromising literalism that galvanized Woodrow's opponents in the evolution trial had long been entrenched in the Southern Presbyterian psyche and was safeguarded for reasons that, as we shall see, were as ideological as they were theological.

Prehistory of the Dispute

In the years preceding the Woodrow public spectacle, internecine skirmishing had regularly broken out over how southern Presbyterians should respond to the new issues modern science was placing on their agenda. At the height of the Woodrow debacle in the mid-1880s, Flinn noted that the majority of those throwing their weight in with Woodrow were comparatively young. Those who actively took up arms against him were of an older generation whose memories extended back to the earliest days of the Perkins Chair.[64] The hostilities that continued to dog Woodrow's efforts echoed the anxieties that had plagued the project from the start. Disquiet over evolution was simply the latest in a sequence of traumas that scientific developments were now regularly inducing.

In 1857 the Synod of Mississippi passed a resolution, on account of "the most insidious attacks . . . upon revealed religion, through the Natural sciences," recommending "the endowment of a professorship of Natural Science as connected with revealed religion in one or more of our theological seminaries."[65] And so it was that in 1859, the Synods of South Carolina, Georgia, and Alabama acted on the benefaction of Judge John Perkins Sr. to establish at the Columbia Seminary "The Perkins Professorship of Natural Science in Connexion with Revelation." Rev. James A. Lyon, the minister of Perkins's congregation, warmly welcomed the move. He thought seminary graduates were woefully equipped to "defend Christianity against the virulent and oft-repeated attacks made upon it through the medium of the sciences." Particularly pernicious was the assault on the doctrine of "the Unity of the races" which was designed to impugn the "Bible account of the origin of man." Now routinely supported by reports "that certain 'monuments,' 'inscriptions,' 'astronomical hieroglyphics,' etc, have been discovered in Egypt, India, or the ruins of Ninevah"—precisely the currency in which polygenist anthropologists routinely traded—this species of skepticism was spreading like wildfire. Young ministers "in some frontier settlement" were in dire need of a solid grounding in the kind of informed Christian anthropology that only an occupant of the Perkins Chair could provide. What was needed was cool-headed, knowledgeable reflection, not the knee-jerk dismissal of the "religious zealot" who "denounced as Infidels and Atheists" the advocates of every new scientific advance.[66]

Richard S. Gladney was also concerned that students should be equipped to "vindicate . . . truths against the assaults of infidelity." Chief among these were works like the radically polygenist *Types of Mankind* that Josiah Nott and George Gliddon had brought out in 1854. At the top of the list of topics that the occupant of the chair must urgently address, Gladney also identified the obsession of contemporary ethnologists to "disprove the origin of men from a single pair." This "infidel" development was as incoherent as it was pernicious, for its "advocates" had not been "able to determine, with all their collections and measurements of craniums and facial angles, their observations on the pelvis and other parts, the color of skin, the weight of brains . . . whether there be two, three, five or a dozen different species." To Gladney, racial diversity was the product of natural forces, and he mused that "*the law of self-development*" refuted "the supposition that there were a multiplicity of origins." Plainly, the current flurry of excitement among certain southern naturalists embracing human polygenesis was of pressing importance in Gladney's mind as he anticipated the establishment of the Perkins Chair. For him, as for Lyon, the problem of racial variety was intimately bound up with the whole question of human origins. Human diversity and human development went hand in hand. But Gladney was also concerned that the endowment should not be "squandered in bricks and mortar, as if these were as important as brains." The sorts of brains he had in mind were not those of narrow scientific specialists but rather the likes of "a McCosh, or a Hitchcock"; what he envisioned for the Perkins professorship, to put it another way, was "not a man of science who has added theology to science, but a theologian, who has called science to his aid."[67]

Perkins himself had originally had Lyon in mind for the post, but the formal appointment initially lay in the hands of the Georgia Synod and was delayed because, in the interim, Woodrow's name had surfaced as a potential candidate. The election was postponed for a year; it was November 1861 before the other synods had recorded their endorsement and Woodrow delivered his inaugural lecture.[68] Even before the Perkins professor uttered a word in his official capacity, doubts about the whole enterprise began to surface. Robert Dabney (1820–1898)—doyen of the southern theologians, college professor, sometime Confederate officer, and Stonewall Jackson's chief-of-staff and biographer—was a social and theological conservative.[69] And he was deeply troubled. He later reminisced that from its inception the Perkins Chair tended "toward naturalistic and anti-Christian opinions."[70] Woodrow suspected that such opposition would come, for he confided to his friend and colleague Thomas Smyth, "I fear that my views, instead of a defence of the truth, might be regarded, by many of those who are foremost in our church, as a surrender of much that is most precious."[71]

In July 1861, Dabney aired his thoughts on geology and the Bible for readers of the *Southern Presbyterian Review*. The first couple of paragraphs set the tone; he registered his "protest against the arrogant and offensive spirit in which geologists have often . . . met clerical criticisms of their reasonings." He was no less perturbed by the species of contemporary harmonizers of Genesis and geology who were forever ensnared by "that shallow and fickle policy" of promoting "some newly coined exposition" of scripture "to suit some exigency of new scientific discovery." Dabney was plainly allergic to novelty: within two sentences he had already castigated "new-fangled" hypotheses, "newly coined" biblical expositions, and "new-fangled explanations."[72] His strategy for dealing with scientific challenges was thus as conservative as it was conventional. No doctrinal reassessment need be contemplated until a scientific claim had been established beyond all doubt. It was a theological all-win strategy, for he so construed scientific demonstration as to rule out *any* rethinking of biblical interpretation in the light of geological science. Simply put, any act of divine creation would produce a world with the appearance of age; so geological "evidence" was no witness to truth, only to presumption.[73] If a scientific observer had been present even a moment after the act of creation, he insisted, "any observations or inferences he might have drawn from the *seeming* marks of the working of natural laws upon them would have been worthless to prove that those specimens originated in natural laws." Therefore, all the harmonizing efforts of such luminaries as Edward Hitchcock,

Thomas Chalmers, and Hugh Miller were premature and unwise. Dabney's tactic was noticeably isolationist but, from his point of view, well worth taking, for it preserved the thing he most cherished: a literal Bible. For this position enabled him to cling to the traditional dogma that "God brought our world out of nothing into an organized state, about six thousand years ago, and in the space of six days."[74] What Dabney gained by this move, as we will see, was something of monumental cultural significance.

On 22 November 1861, Woodrow set out his wares in his inaugural address as the Perkins professor. After laying out the conditions of the benefaction, the nature of the post, and the like, he turned to the issues that currently disturbed religious sensibilities and that he regarded as his "chief duty to examine": the age of the Earth, the extent of the Noachian flood, and the unity of the human race. His tactics were fundamentally at odds with Dabney's. The "most untrammeled freedom of inquiry" had to be extended to those investigating such questions, as well as "the most unbiased readiness to accept as truth whatever is proved." Scientific investigation had to remain exempt from ecclesiastical prescription. As for his stance on a number of specific issues, he made it clear that the antiquity of the Earth and gradualism in the history of life were "certainly true" but that "the teachings of such ethnologists as deny the specific unity of the human family" were "certainly false, or at least wholly unproven." In concluding his lecture, Woodrow warned of two particularly destructive attitudes that tended to afflict the whole science-faith arena. On the one hand, there was a need "to be continually on our guard against a dogmatic adherence to opinions which may not be well founded, and the denunciation as infidel of whatever differs from our own." On the other hand, there was "a facile acceptance of every novel and attractive hypothesis which may spring up in the field of science." Neither approach was designed to promote genuine harmony between science and religion. But perhaps the greatest danger of the hour was the survival of "the old spirit" that "would crush all progress in science, if such progress disturb, in the least, cherished views which may be without real foundation in the Bible, by the employment, not now of material instruments of torture, but by that which has with too much truth been denominated '*odium theologicum.*'"[75]

By now, of course, the cultural politics of the Perkins Chair were overshadowed by political events on an infinitely grander scale. It was nearly a year since the South had declared its secession from the Union and Fort Sumter in Charleston harbor had been bombarded in April 1861. Woodrow joined other southern patriots in supporting the Confederate cause by maintaining a medical and chemical Confederate laboratory and serving as chaplain. In December, the denomination split, the new Southern Presbyterian Church came into being, and the seminary closed its doors to students for the duration of the conflict.

Still, Woodrow did find time amid the horrors of the Civil War to turn his mind to matters of science and religion. Evidently bristling from Dabney's recent assault, he put pen to paper in 1863 to defend geology from such adversaries.[76] When he identified as a first-order "mistake of anti-geologists" that they "suppose it [geology] to be cosmogony," it is hard to believe he did not have in mind Dabney's assertion that geology "is virtually a theory of cosmogony."[77] It was surely the same when he railed against those "exciting suspicion and prejudice against the geologist, by raising the hue and cry of 'rationalist,' 'skeptic,' 'infidel,' 'atheist.'"[78] But he had other targets in mind, too, not least the scriptural geologists David and Eleazar Lord, whose views had been the subject of a recent piece in the *Southern Presbyterian Review*, and their British counterparts Granville Penn and George Fairholme.[79] It profoundly troubled him that the antigeology faction among the Southern Presbyterians was doing all in its power—whether by accident or design—to isolate itself from the theological mainstream. By itemizing some of the luminaries in the pantheon of theological allies of geology—Thomas Chalmers, John Pye Smith, David King, John Harris, Hugh Miller, Edward Hitchcock, Adam Sedgwick, William Buckland, never mind a host of others beyond the English-speaking world—Woodrow hoped to expose the "mistake of anti-geologists, who are so fond of classing geologists with infidels, or with those who know little of the Bible and its teachings."[80]

Another ten years would elapse before Woodrow again put his thoughts on the whole subject in print. In the meantime, his warnings were falling on deaf ears. Over the following decade or so, Dabney kept up the pressure, becoming increasingly fierce in his denunciation of those seeking to harmonize Genesis and geology. In a memorial entitled "Theological Education" in 1869, for example, he turned his guns on those championing a role for science in a seminary curriculum. The whole idea was "extremely objectionable." "The tendencies of such a course," he announced, "will be mischievous" for teacher and pupil alike. Visions of students "babbling the language of geology and ethnology with a great deal more zest than they recite their catechisms" alarmed him. Worse still, it would take several generations before the full force of the "evil comes to a climax." Why? Because the very fabric of "these sciences is essentially infidel and rationalistic; they are arrayed, in all their phases, on the side of scepticism." Now, in language more brutal than ever, he lambasted Hugh Miller's accommodationist *Testimony of the Rocks* as "thoroughly impregnated with the secret *virus* of rationalistic infidelity."[81]

Dabney was now at full tilt. In 1871 he issued "A Caution against Anti-Christian Science," which had begun life as an address to the Virginia Synod that October. Again he reiterated the charge that "physical science always has some tendency to become anti-theological" and "to exalt naturalism." "Positivism," he declared, was the "vain, deceitful philosophy" of the day and was animated by a passion to undermine the detailed authority of scripture. Those who embarked on adjusting their understanding of the Bible to meet its challenges had already sacrificed the plain literal meaning of the six days of creation, localized Noah's universal flood, and surrendered the Mosaic chronology to shaky claims that "man has been living upon the globe . . . for more than twenty thousand years." Now aggressive ethnologists were asserting that "there are several distinct species" of human "having different origins," and evolutionists were insisting on the "monstrous idea, that the wondrous creature, man, . . . is but the descendant . . . of a mollusc or a tadpole."[82]

What is notable in Dabney's repeated litany of scientific profanities was the close association he perceived between the antiquity of the human race, racial polygenism, and human evolution. The two figures he cited as illustrating good practice were John Bachman and James L. Cabell, both southerners, who had defended the Christian doctrine of the unity of the human race "against the assaults of natural science, with the weapons of natural science." Dabney's choice of exemplars indicates what he thought was the most critical arena in which the relationship between science and theology played out.[83] And so, in language befitting a Civil War veteran, he declared open war on the entire enterprise. "You must resist," he declared, "or you must practically surrender your Bibles. You will have to 'take sides' for or against your God." It was a battle cry to take up arms against the kind of gross materialism witnessed a month or two earlier at the Indianapolis meeting of the American Association for the Advancement of Science where, Dabney reported, "the great majority of the members from the Northern States openly or tacitly disclaimed inspiration."[84] It sounded like a call to reengage in civil war by other means.

By 1873, Woodrow had had enough. In a fifty-page rejoinder entitled "An Examination of Certain Recent Assaults on Physical Science," he explained why. Dabney's perverse inclination to catch a whiff of unbelieving rationalism in any rapprochement with modern scientific inquiry had routinely surfaced in assembly memorials, lecture courses to his students in Union Theological Seminary, published curricula, and printed sermons, Woodrow complained. Talk of infidel science, atheistic geology, sensualistic philosophy, and the like just came too easily and insidiously to Dabney's tongue. It was horrifying to see how he had been "sounding forth the alarm" and stirring up fellow church members "to rise in arms against Physical Science as the mortal enemy of all the Christian holds dear, and to take no rest until this infidel and atheistic foe has been utterly destroyed." Such tactics were frankly "dangerous."[85] But they were also foolish. Dabney failed to see that the idea that humanity's fall from grace undercut the reliability of the claims of natural science was self-refuting, for the consequences of original sin must apply no less to theological declarations. Dabney's distrust of the deliverances of human reason, Woodrow insisted, had profoundly cynical implications, for it committed him to a "cheerless scepticism" about the use of reason *tout court.* Dabney's tactic of "nicknaming physical science 'vain, deceitful philosophy,'" his persistent habit of claiming that science was intrinsically naturalistic, and his allegation that the "spirit" of the natural sciences was "essentially infidel and rationalistic" all rubbed the grain of Woodrow's professional instincts the wrong way. More specifically, Woodrow set about addressing one by one the "encroachments," as Dabney had dubbed them, that science was making on scripture.[86] Underlying his response to each of these particulars was his conviction that Dabney's efforts to subject scientific inquiry to ecclesiastical regulation were profoundly misguided. "The truth is," Woodrow announced, "that natural science is neither Christian nor anti-Christian, neither theistic nor atheistic, any more than the multiplication table. When we can speak of a Christian law of gravitation, or an infidel law of definite proportions, or a rationalistic order of succession in the strata . . . then we shall be able to speak of Christian and atheistic natural science, and not until then."[87]

Relationships were now deteriorating rapidly, and the language became increasingly tetchy. In October Dabney returned fire. Claiming that Woodrow had misread his intentions and misunderstood his arguments, he reiterated his case. After sketching in the background to the whole business—not least his own concern to get the seminary on a right footing at a time when the South was "passing anew through a formative state"—he again portrayed the "study of modern geology" as "seductive" and "naturalistic" in spirit. Unlike "all other human sciences, as law, chemistry, agriculture," subjects like geology were essentially skeptical because they came "into competition with the theistic solution of the question of the origin of things." Along the way Dabney raked over the ashes of some earlier theological squabbles and engaged in a bit of skirmishing over stratigraphical labels. But nothing had fundamentally changed, save for the tone, which was darker. Beneath a veneer of esteem for Woodrow's "sound and safe science" were more caustic sentiments that bubbled up to the surface from time to time. Woodrow— with his "fifty-two dreary pages of criticism"—had launched "out into the most amazing misunderstanding and contradictions" and taken pleasure in meting out "slashing criticism of one who had given no provocation to him." Had he "learned manners in the school of Dr. Woodrow," Dabney barked, he was sure he would be "warranted in retorting some of his very polite language" to show that Woodrow knew just about as much psychology as a wayward college boy.[88] And so it went.

The following April, in 1874, Woodrow snapped back. For seven or eight pages he danced around the target, recalling earlier theological opposition to humanist learning, reminiscing on Protestant and Catholic hostility to Galileo, and reflecting on more general religious railing against natural science. Soon, however, his real quarry came into full view: Dabney, "a gentleman who for talent and zeal and earnestness and many estimable qualities deserves to be highly honored by all who know him." Interrogating his text with the precision of a neurosurgeon, Woodrow exposed contradictions, misunderstandings, and confusions and printed in full those parts of a private correspondence that Dabney had omitted from his previous statement. Was it not patently self-contradictory for Dabney to insist that seminaries should only teach the denomination's settled theology when he himself gave instruction in that most positivistic of enterprises, mental science? Again, surface reverence eventually conceded to more acerbic sneers, and as Woodrow worked up to his final crescendo, he depicted Dabney "with a shout of triumphant laughter" as "furiously brandishing his mop against each succeeding wave [of science], pushing it back with all his might." It was, to Woodrow, a pathetic gesture. He went on:

> But the ocean rolls on, and never minds him; science is utterly unconscious of his opposition. If this were all, the contest would be simply amusing. But it is not all . . . There are numbers, even among our most learned and devoted ministers, who share these views which we regard as so inconsistent with the truth and as so fatal in their consequences. We would fain do something to prevent these terrible consequences by persuading all whom we can influence to review the ground on which they base their present opinions; confident that a fair reëxamination will without fail lead to a change of mind.[89]

And there, for the meantime, matters rested. The evolution controversy that blew up with the impeachment of Woodrow in the mid-1880s was clearly not an isolated incident in the Southern Presbyterian encounter with science. Indeed, many of the crucial elements that surfaced in that public spectacle had already been aired in the scuffles surrounding the establishment of the original Perkins Chair and in the increasingly testy squabble between Woodrow and Dabney more than a decade before the Columbia heresy trial. Dabney was shaken to the core by fears about the skeptical atmosphere in which the natural sciences—geology in particular—conducted their affairs, and the reckless way in which the straightforward, literal sense of scripture was sacrificed on the altar of scientific respectability. To him, the dangers were plain for all to see. Woodrow, by contrast, could only see ecclesiastical isolationism, philosophical myopia, and a pathological struggle to exert control over theological education in the whole business. This was the cultural context into which Darwin and his theory were soon to be pitched. But to appreciate the fullest depths of what Dabney and his coterie felt was at stake, we need to attend to a number of concurrent developments over the following few years which serve to expose even deeper historical and psychological roots of the resistance.

Toward a Deeper Archaeology

In the months that followed the clash between Woodrow and Dabney in the pages of the *Southern Presbyterian Review,* one or two other voices were to be heard on the subject. In the January and October 1874 issues, for example, the aging Richard Trapier Brumby (1804–1875), successively professor of geology, mineralogy, and other sciences at the University of Alabama, the University of Georgia, and South Carolina College, took up the topic. His first piece was an extended review of a couple of recently published works, but he used the opportunity to reflect more generally on the state of the union between science and religion. He himself was certain that Darwin was one of that class of "infidel speculators" who mobilized science for subversive purposes. But it bothered him that theologians, however expert, too rashly entered the field of combat and did little more than "aid sceptical writers" by disseminating the false notion that science and religion were necessarily at odds. The "overthrow" of the irresponsible speculations of skeptical science should be left to "scientists alone." "Few men are qualified to write wisely on any supposed discrepancy between science and revealed truth," he insisted. "To discuss any such question, so as not to excite in the public mind the latent tendency to unbelief, requires more knowledge and wisdom than most men—even learned theologians—possess." When he further proclaimed that the sciences of "chemistry and geology" were "great systems of truth, to which the human mind cannot refuse assent"—however much they had been hijacked by skeptics—it was not difficult to see just who he had in mind.[90]

In October, Brumby took up the theme of gradualism in scientific knowledge, scriptural revelation, and the natural world. The whole piece was designed to promote the idea of progressivism and thereby undercut a too-ready resort to divine interventionism on the part of some critics. Along the way he paused to confirm his view that "evolution by *transmutation*, by selection" was indeed the province of "sceptical writers," but nonetheless he held out the idea of progressive creationism—namely the extinction of species and their "substitution by direct creation . . . of higher types." To him this was a form of "evolution" that was simply a "fact." As for the human species, he suggested that even if "the original creation of several species" of humans—polygenism—were to be established, he was sure that it "would leave unshaken . . . the belief that Adam was divinely made the psychological and representative head of all human creatures." He even went so far as to speculate that the Bible implied "that the descendants of the highest, representative, Adamic race will eventually subdue and exterminate inferior races, and occupy the whole earth. The almost total disappearance of American Indians—the work of a few centuries—and many other facts favor this view."[91]

In the meantime, a much younger writer, William Smith Bean (1849–1920) of Washington, Georgia, delivered an analysis of the modern scientific enterprise, noting how figures like Huxley and Winwood Reade were leading their followers "into dismal and chilling regions." Rather than acquiescing to despair, he cautioned his readers to consider whether such conjectures were really intrinsic to scientific practice itself. "The one mistake, constantly made in this feverish, speculative age both by theologians and men of science," he mused, "is *haste*." Instead of wallowing in idle speculation, scientific practitioners should stick to facts, facts, and yet more facts. Theologians, often no better than "half-informed" meddlers, called to Bean's mind the "monkish suspicion and mediaeval narrowness" that had afflicted the church for too long. "They gaze on the spectroscope, the microscope, and the scalpel," he went on, "with the horror of some mitred bishop or abbot contemplating the retorts and crucibles of alchemy, or the crabbed formularies of the black art. The evil of such opposition is, that it begets or fosters a spirit of scepticism." Their "sneering tone of ridicule" did nobody any good; it only succeeded in alienating those fellow believers who looked respectfully on the achievements of modern scientific inquiry.[92]

However conciliatory to modern science these essayists were, their voices were largely drowned out of the denominational conversation by the thundering force of Dabney's censure. The following year, his lengthy analysis, *The Sensualistic Philosophy of the Nineteenth Century Considered* (1876), made its appearance. It continued—often using the same scornful language—the campaign of denigration on which he had been embarked for at least a decade and a half. His target was the naturalizing tradition of scientific inquiry and social philosophy which sought to reduce mind and spirit to mere matter in motion. Its rationalistic ethos, he was convinced, was designed to foster skepticism, to obliterate morality, and to extinguish all belief in the supernatural. Positivism, as he dubbed this alien outlook, was afflicting many groves of scientific and philosophical investigation by reducing everything to sheer sensation. To Dabney, positivist philosophy, the horrors of the French Revolution, the specter of atheism, and the evils of free love were all netted together. And this was the company that Darwinian evolution kept. So, alongside his defense of the coherence of a priori reasoning and his critical analysis of the philosophies of Thomas Hobbes, John Locke, Étienne Bonnot de Condillac, and their latter-day materialist disciples, he devoted two substantial chapters to an assessment of Darwinism itself.

In the first of these evaluations, Dabney explained that the kernel of the transmutation hypothesis lay in its naturalizing impulse, its obsession with finding a comprehensive naturalistic explanation for nature's variegated forms. Nowadays, it was championed by Darwin, but the fancy had earlier been promulgated by the likes of Democritus, Lord Monboddo, and Lamarck. Blind natural selection and the survival of the fittest were only the latest mechanisms called upon to establish "the process of evolution as entirely unintelligent." Nor did the champions of these doctrines stop short at physiological modification; human morality was also brought within the embrace of evolutionary imperialism. Tyndall's Belfast Address, Dabney noted, had been nothing less than a reckless folly "to revive the forgotten system of Democritus." Yet for all these absurdities, to Dabney they paled into insignificance beside another conundrum attending the reception of Spencerian evolution. "One of the most astounding things connected with this monstrous aggregation of confusions and assumptions," he declared, "is the applause it has received from some critics professedly Christian."[93]

Dabney's second attack dwelt on evidence that, he was certain, refuted the Darwinian scheme. Here some of the concerns closest to his heart conspicuously manifested themselves. He was particularly persuaded that, in the state of nature, reversion to type—as routinely witnessed by cattle breeders and bird fanciers— militated against long-term species variation. Only "the most artful vigilance" could prevent "heterogeneous individuals from propagating" and corrupting "the combined type which is required." Under natural conditions, by contrast, variation produced "a more confused and degraded progeny." Nowhere was this more dramatically disclosed, Dabney maintained, "than in the human species, where the savage anarchy produced by the violences of the stronger is always found to reduce the whole tribe to destitution." The racial implications were plain: "Why else is it," Dabney asked, "that Bushmen are poorer, shorter, uglier and feebler than Englishmen?" The creation had fixed bounds, established hierarchies, set limits, as hybrid sterility amply attested. "Providence," he told his readers, had installed mechanisms to prevent "that disastrous intermingling of types of organization, shading off in every direction into interminable confusions." Darwin's theory challenged Dabney's cozy cosmos head-on. The stability of the entire natural, moral, and social order was under the direst threat. No wonder Dabney proclaimed that the "conclusions of Evolutionism . . . are an outrage to the manhood of our race. What foul, juggling fiend has possessed any cultivated man of this Christian age, that he should grovel through so many gross sophistries, in order to dig his way down to this loathsome degradation?" The whole Darwinian enterprise was nothing but a "pretext for materialism, sensuality, and godlessness."[94] How different that was from the old culture of southern Christian civility. As he put it in a concluding rhetorical flourish: "The fair mode of comparing the fruits is to contrast the whole body of atheistic materialists with the whole body of sincere Christians. Then, on the one side we have such characters as the Jacobins and the *sans-culottes* of Paris in her two reigns of terror, and those original 'Positivists,' the Bushmen of Africa, and the blacks of Australia; on the other, we have nearly all that has been good and true and pure in Christendom and without it."[95]

During 1876, the year when his *Sensualistic Philosophy* first saw the light of day, matters of race, politics, and social fixity were dominating Dabney's mind for other reasons, too. During that decade, efforts were being strenuously made to extend to the American South the common school system of free education in which all children would receive instruction funded by local taxes. As a long-standing apologist for the Lost Cause of the Confederacy—a social and cultural movement dedicated to the retrieval of southern chivalry and nobility in the face of northern decadence and immorality—Dabney was profoundly perturbed by this newfangled piece of northern meddling.[96] Not only was it politically repressive; it was antiscriptural through and through. In February, then, he submitted an article to *The Southern Planter & Farmer* on the touchy topic—"The Negro and the Common School."

In mourning for the old South's fading age of deference, Dabney announced up front that the idea that black children should mingle with their white counterparts in a common school was further evidence of "the falsehood and deadly tendencies of the Yankee theory of popular State education." Excoriating language flowed from his pen: the whole idea was a "Yankee heresy," "rank," "fatal," "deceptive, farcical and dishonest."[97] It galled him to think that taxes from "white brethren" would be diverted "to give a pretended education to the brats of the black paupers, who are loafing around their plantations." The whole idea went against nature itself. For what made "the negroes, as a body . . . glaringly unfit" for education and suffrage alike was a disabling litany of inferiorities: the "inexorable barrier of alien race, color, and natural character . . . a dense ignorance of the rights and duties of citizenship . . . a general moral grade so deplorably low as to permit their being driven or bought like a herd of sheep . . . a parasitical servility and dependency on nature . . . an obstinate set of false traditions." What the entire northern project to subvert the Christian civilization of the Old South completely failed to grasp was that in "every civilized country, there must be a labouring class," and Dabney could "not see any humanity in taking the negro out of the place for which nature has fitted him." That went against the natural order itself . . . as did that other "abhorred fate, *amalgamation*." With a pathological fear of hybridity in every shape and form,[98] Dabney was certain that integrated education would unleash a "curse of mixed blood"—"this poison of hybrid and corrupted blood"—which would "complete the destruction of the white States."[99] When he spoke to the November 1867 meeting of the Synod of Virginia on the subject, his voice reportedly "trembled with emotion, his frame shook, his eyes snapped fire."[100] For to Dabney, the consequences would be calamitous:

> The satanic artificers of our subjugation well knew the work which they designed to perpetrate: it is to mingle that blood which flowed in the veins of our Washingtons, Lees, and Jacksons, and which consecrated the battle fields of the Confederacy, with this sordid, alien taint, that the bastard stream shall never again throb with independence enough to make a tyrant tremble. These men were taught by the instincts of their envy and malignity, but too infallibly, how the accursed work was to be done. They knew that political equality would prepare the way for social equality, and that, again, for amalgamation.[101]

Such segregationist refrains reverberated around Dabney's mind during the following months when he engaged in a series of newspaper exchanges in the *Richmond Dispatch* and the *Richmond Enquirer* on the common school question with W. H. Ruffner, superintendent of state schools.[102] Dabney reiterated his objections to the whole system but made it clear that besides its unnatural character, it also sought to undermine "that providential order which God has imposed upon society." Accordingly, common schools were revolutionary in spirit. The idea that the "children of the Commonwealth are the charge of the Commonwealth is a pagan one," he declared, and stood in total opposition to "Divine Providence," which "determines the social grade and the culture of children." The system thus fostered "a universal discontent with the allotments of Providence, and the inevitable graduations of rank, possessions and privilege" and failed to acknowledge that "God has made a social sub-soil to the top-soil, a social foundation in the dust, for the superstructure." Armed with this social theodicy, Dabney again resorted to vicious language in his resistance to what he called this "ruthless, levelling idol." The thought that in "order to receive the shallow *modicum* of letters" dispensed in common schools, white children "must daily be brought into personal contact with the cutaneous and other diseases, the vermin (Yes, dear reader, it is disgusting! . . .)—the obscenity, the profanity, the grovelling sentiments, the violence of the *gamins*" was too great to bear. Any thought that "blacks" were "equal, socially and politically, to the most reputable whites" went against everything Dabney stood for—the arrangements of divine providence, a fixed social hierarchy, and the natural order itself.[103]

This reactionary outlook, of course, was nothing new. For decades, Dabney had been defending the South's social conservatism as grounded in Christianity. Most conspicuously, he had laid out the biblical foundations of the South's slave culture in a 350-odd page apologia for the Lost Cause, *The Defence of Virginia*, which appeared in 1867 but had been completed in manuscript since 1863.[104] From the outset, he made it clear that "abolitionism" was thriving within the church, "more rampant and mischievous than ever, as infidelity." In the State, it continued to flourish "as Jacobinism"—"a fell spirit which is the destroyer of every hope of just government and Christian order." Since he was certain that the "teachings of Abolitionism [were] of rationalistic origin, of infidel tendency, and only sustained by reckless and licentious perversions of the meaning of the Sacred text," he devoted over a hundred pages to culling the Old and New Testaments for every single passage referring to slavery in order to show that it enjoyed the support of scripture. To Dabney, the "scriptural argument for the righteousness of slavery" was irresistible.[105] The life of Abraham, the Mosaic Law, the Decalogue, the gospels, and the Pauline epistles all bore witness to the morality of domestic servitude. Besides, scripture was the only basis on which the institution could be regulated on principles that would prevent its abuse by corrupt slave-masters. In the Bible could be found just the sort of benevolent paternalism on which southern greatness had been built.[106]

He was fully aware, of course, that polygenist anthropology could deliver a scientific rationale for racial hierarchy, but he would have no truck with that repugnant course. "It is not our purpose to rest our defence on an assumption of a diversity of race, which is contradicted both by natural history and by the Scripture, declaring that 'God hath made of one blood all nations of men.' "[107] Righteous slave owners did not need to resort to infidel science to explain themselves; the Bible when read in a straightforward, unfanciful way provided the only grounding that was needed for justifying slaveholding as a divinely ordained institution.

A plain, unadorned, literal Bible was the foundation of southern Christian culture. By the same token, Dabney was no less certain that *sound* science was also on his side.

> The African has become, according to a well-known law of natural history, by the manifold influences of the ages, a different, fixed *species* of the race, separated from the white man by traits bodily, mental and moral, almost as rigid and permanent as those of *genus*. Hence the offspring of an amalgamation must be a hybrid race, stamped with all the feebleness of the hybrid, and incapable of the career of civilization and glory as an independent race. And this apparently is the destiny which our conquerors have in view. If indeed they can mix the blood of the heroes of Manassas with this vile stream from the fens of Africa, then they will never again have occasion to tremble before the righteous resistance of Virginian freemen; but will have a race supple and vile enough to fill that position of political subjection, which they desire to fix on the South.[108]

By this move Dabney was able to retain his belief in profound differences between the races, biblical monogenism by universal descent from Adam, and a rigid racial hierarchy.[109] It was a long-standing coalition among Southern Presbyterians, as well as other denominations.[110] Nearly twenty years earlier, for example, just this conceptual confederation undergirded Thomas Smyth's well-known 1850 attack on Agassiz-style polygenism, *The Unity of the Human Race.* Smyth—a Scots-Irish immigrant, Charleston clergyman, and slave reformer who defended the institution on biblical grounds—was deeply troubled. To him the polygenetic "theory would be very inexpedient and suicidal to the South in the maintenance of her true relations to her colored population." He thus marshaled scientific and scriptural arguments, bolstered with historical, philosophical, and linguistic findings, to undermine it. Indeed, he quarreled with the American School of Anthropology, championed by polygenists like Samuel George Morton, Josiah Nott, and George Gliddon, in part because he felt their undertaking would undermine the traditional biblical basis of southern slavocracy: "The introduction in the South . . . of this novel theory of the diversity of races, would be a declaration to the world that its institutions could no longer rest upon the basis which has always been hitherto assumed, and that this theory has been adopted for mere proud, selfish, and self-aggrandizing purposes. This theory is further impolitic to the South, because of its immoral, anti-social, and disorganizing tendencies. It would remove from both master and servant the strongest bonds by which they are united to each other in mutually beneficial relations."[111]

Put simply, polygenist anthropology bestialized slavery; biblical theology sanctified it. For Smyth, slavery had the benediction of scripture, and any undermining of biblical authority was thus socially subversive. For generations, polygenism's spiritual infidelity with its reckless attack on the Mosaic record was well known, but now its seditious politics were laid bare. Christian slavery, to Smyth, did not mean that slaves were chattels; it gave rights and privileges, it was providentially ordered, but it was slavery all the same. No wonder polygenism was "impolitic to the South."[112] It would wreck the building blocks from which southern civilization had been raised.

In the same year, George Howe (1802–1883), professor of biblical literature at the Columbia Seminary, took exactly the same line. In a review of Josiah Nott's *Two Lectures on the Connection between Biblical and Physical History of Man,* a work Howe read with "painful astonishment," he urged that the polygenetic "hypothesis" was contrary *"to every declaration of the Scriptures."* To Howe the consanguinity of the human race was scripturally sanctioned and scientifically established, a claim he substantiated by quoting lengthy passages from Bachman's soon-to-be-published *Unity of the Human Race.* He suspected that Nott's "desire to find a new basis for slavery" had induced him to seek "to destroy the credit of the Scriptures," for after all its "plain, unperverted sense . . . is, that all men, whatever be their type and complexion proceeded from one stock, from one and the same progenitors." The radicalism of "new and frisky sciences" like ethnology and the "perversions of Abolitionists" stood together; both overthrew biblical revelation and traditional authority. Howe thus abominated efforts to justify slavery in the anthropological currency of cranial capacity, skin color, or hair texture, preferring instead the frank and simple teaching of the Bible, under which "the slaveholding patriarch, the slaveholding disciple of Moses, and the slaveholding Christian lived, protected and unrebuked." Indeed, he confessed that his "greatest anxiety at present is, as to the effect these denials of the Unity of the Race, in the very face of the word of God, will have upon the institution of slavery, in which the entire prosperity of the Southern States is at present bound up."[113]

Howe's credo dovetailed neatly with Dabney's outlook. As we have seen, he had long enthused over the scientific contributions of southern monogenists like Bachman and Cabell who had marshaled "true" science against the perverse anthropology of the Charleston naturalists. Now the latest scientific perfidy, the theory of evolution—whether of Darwinian or Woodrow stripe—challenged the foundation of scripture, slavery, sound science, and social stratification on which southern civilization rested. Constitutionally resistant to theories that played metaphorical with scripture, he considered Darwinism as simply the latest expression of that infidel specter that hounded Christian culture from generation to generation.

As he explained in his 1888 review of "Anti-Biblical Theories of Rights," when "the friends of the Bible win a victory over one phase of infidelity," the forces of irreligion mounted an offensive from some other quarter. First it was the "rational deism" of the likes of David Hume. Then "biblical criticism in the hands of Bengels, Delitzsches" and others of their stripe had their day. Now it was "professed social science" with its "dogmas of social rights which are historically known as Jacobinical, and which have been transferred from the atheistic French radical to the free Protestant countries."

By contrast, biblical governance revealed "God's preference for the representative republic as distinguished from the levelling democracy" to which the South was now subjected. Biblical equality, Dabney insisted, was "moral," not "mechanical," and most assuredly acknowledged "the relation of superior and inferior." For these reasons, he explained, the Southern Presbyterian Church had always subjected "any science so-called, whether psychological, moral, or even physical" to the test of scripture. That was why the General Assembly had dealt as it did with Woodrow's approach to evolution: "So, recently, our Assembly, upon perceiving that a doctrine of mere physical science, evolution, was liable to be used for impugning the testimony of Scripture, dealt with that foreign doctrine both didactically and judicially. They were consistent. For, I repeat, whenever any doctrine from any whither is employed to assail that divine testimony . . . there the defensive discussion of that doctrine has become theological, and is an obligatory part of the church's divine testimony." For Dabney, sacrificing a literal reading of the Bible's account of genesis, in general, and "of the sin and fall of the race in Adam," in particular, to the demands of evolutionary science would be to surrender everything southern culture had stood for in resisting the treacherous forces of infidel northern Jacobins. For all the "rationalistic glozings of deceitful exegesis," it ultimately boiled down to this: "He who attacks the inspiration of Moses attacks also the inspiration and moral character of Jesus." There was only one safe path for the remaining faithful—"stand on the whole Scripture, and refuse to concede a single point."[114]

Dabney was certainly not alone in his reading of evolution as posing the direst of threats to southern interests. By the mid-1880s, another of Woodrow's opponents, George Armstrong, had emerged, according to Ronald Numbers, as "southern Presbyterianism's leading voice on matters of science and religion."[115] In 1886, the year Armstrong's anti-Woodrow report on the immediate creation of Adam and Eve was adopted by the Augusta Assembly, his *Two Books of Nature and Revelation Collated* made its appearance. In more than two hundred pages, he defended the biblical view of the world over against modern science in various guises. The substance of the two lectures he had presented back in 1884 were the backbone of the book, but he supplemented these with further critical observations on "primeval man" and the reliability of the Mosaic cosmogony, a critique of higher textual criticism, and a refutation of Huxley and Tyndall on the subjects of providence and prayer.

Two things stand out in his diagnosis. First, Armstrong's thinking about evolution was comprehensively shaped by its implications for the human race. Right from the start, he engaged with Nott and Gliddon's *Types of Mankind* and physical anthropology more generally, and he persistently turned to writers concerned with the unity and diversity of the human race—Bachman, Cabell, De Quatrefages, and the like—to establish the permanence of species and the profound irreconcilability of evolution with "the account of man's creation given in Scripture."[116] Human unity, he believed, had ultimately been confirmed by the best science, and this endorsement bore witness to the wisdom of holding steadfastly to the biblical account instead of hastily embarking on desperately strained harmonizing strategies.

Second, his profound distaste of modern biblical scholarship was all of a piece with his aversion to evolution. In his essay on the Pentateuch, for example, his targets were William Robertson Smith and, more particularly, his American counterpart, the brilliant Hebrew and Semitic linguist Crawford Howell Toy of Harvard, who had grown up in Armstrong's own city—Norfolk, Virginia. To be sure Armstrong presented what he considered hard evidence in support of the Mosaic authorship of the Pentateuch over against Toy's wild speculations, but it troubled him that such higher critics assumed "the truth of the hypothesis of evolution." The direction in which their espousal of evolution led was not a happy one. "In common with the advocates of the theory of the evolution of man from the brute," he observed, "Dr. Toy . . . assumes that man, as man, began his course upon the earth as the most ignorant, debased, and superstitious savage; and gradually, by his own efforts continued through ages, worked out a civilization and a religion for himself, that God having created him—if, indeed, He did create him, a pitiable troglodyte, like the Digger Indians of the West—left him to work out his destiny as best he could." How different that picture of early Israelite polytheism and fetishism was from the "authentic history" of Moses.[117] That Toy lapsed into Unitarianism around 1888 only confirmed his critics' conviction that their attacks on him were justified.

Like Dabney, the roots of Armstrong's investment in the detailed accuracy of the Mosaic record—and the Bible more generally—long predated the Woodrow debacle over Darwinism. A veteran proslavery apologist, Armstrong had published *The*

Christian Doctrine of Slavery (1857), in which he too claimed that abolitionism had sprouted from the infidel philosophy that had stimulated the French Revolution and emphatically not from the biblical account of civil government and social hierarchy. At a time when polygenist scientists in the South were calculating brain weights, computing cranial capacity, cataloguing skin hue, and otherwise doing all in their power to establish a race hierarchy that confirmed a separate origin for each human race—a theory that would undermine the authority of Moses—Armstrong found in a "faithful exhibition" of scripture all the justification the South needed for its racial politics. That slavery nowhere appeared in any scriptural catalogue of sin, that good slave masters featured positively in various parables, that the image of the bondman was regularly used to depict authentic discipleship, that slaveholders were bona fide members of the church in apostolic times, and that various texts set out appropriate reciprocal duties between masters and slaves—all this biblical evidence was marshaled in support of the institution. No less significant was what he took to be Pauline endorsement of detailed Levitical regulations on slaves as heritable property together with the fact that the authorization of slavery in the Mosaic law more generally was never abrogated. Not surprisingly, in his review of the sacred history of slaveholding he grounded the entire subject in the provisions that Moses had laid out for the institution in the Pentateuch. For all these reasons, Armstrong considered anyone—and he had in mind Albert Barnes, author of *The Church and Slavery*—who thought otherwise an "infidel." "You cannot enter my pulpit to preach on slavery," he announced, "for the same reason that the Free-lover cannot enter it to preach on marriage, or the Socialist to preach on the relation of parent and child." Whereas polygenism dehumanized Africans, biblical slavery dignified them. As he put it in his concluding remarks, "The African slave, in our Southern States, may be deeply degraded; the debasing effects of generations of sin may, at first sight, seem to have almost obliterated his humanity, yet is he an immortal creature; one for whom God the Son died; one whom God the spirit can re-fashion, so as to make him a worthy worshipper among God's people on earth."[118]

Another vocal antagonist in the Woodrow affair, John Girardeau, who, we recall, had made it clear in the debate that the immediate creation of Adam from literal dust was a non-negotiable doctrine, had long defended the South as a fundamentally biblicist culture and stood against any downplaying of the literal truth of the Bible.[119] To him, sacrificing the detailed accuracy of any part of scripture meant surrendering it all. Back in 1879, for example, he had written: "If Moses uttered falsehoods, then why not David, and Paul? It does not do to say that Moses was not inspired to teach science; the point is that he was inspired not to teach falsehood."[120] Renegotiating the plain teaching of Moses or tinkering with the denomination's traditional practices were just the thin end of the wedge. He reiterated the same thing during the Woodrow affair: "One thing leads to another. If one exception to the Standards be allowed in an official teacher, another and another may be. Where shall the line be drawn—the limit fixed?"[121]

Now, in the shadow of the Woodrow affair, he raised a troubling suspicion to James Martin, the medical doctor who had defended Woodrow's cause: "May it be that Dr. Woodrow holds . . . to the hypothesis of *Pre-Adamite man*?"[122] This, Girardeau suspected, was one way in which Woodrow—and Martin speaking for him—might preserve the idea that the "first man's body was evolved," while the production of Adam involved divine intervention. Girardeau ventured to hope that this was not the case. For his part Martin made it clear that this resolution was certainly not his or Woodrow's strategy, for both believed that the first true man *was* none other than Adam. Still, Martin did ponder whether the Bible contradicted the idea of pre-Adamite humans. As with the vexed question of other worlds being inhabited by "rational animals," Martin reflected: "Why should Post-Adamite man, or another race of rational animals on this planet, or in other worlds, be less contradictory of the Bible than Pre-Adamite man? . . . Ask the Bible, Are the stars inhabited? It answers not. Ask the Bible, Were there any Pre-Adamite men? Still it answers not."[123]

Such talk could only have traumatized Girardeau. About twenty years earlier, in 1867, a scurrilous volume, *The Negro: What is his Ethnological Status?*, had appeared. Published under the pseudonym "Ariel," its author, Buckner H. Payne, a Methodist clergyman in Nashville, declared that the black races were a pre-Adamite stock and therefore that "the negro is not a *human* being—not being of Adam's race."[124] Comparable ideas circulated in various guises at the time, but Ariel's intervention sparked a heated controversy, not least among Methodists, many of whom were repulsed by the degrading proposal that African Americans were not human.[125] This form of polygenism only confirmed in many minds the evils of ethnological inquiries in the hands of pagan publicists. Girardeau had got wind of the Ariel business and spoke on the subject on Sunday evening, 17 November 1867, in Charleston's Zion Presbyterian Church, condemning it as a foul fanaticism that would deny the entire moral, mental, and spiritual nature of the black race.[126]

Besides, in 1878, the distinguished geologist Alexander Winchell was dismissed from his position at the Methodist Vanderbilt University on account of his scientific views. His acceptance of evolution was a critical factor, but so too was his advocacy of the existence of pre-Adamite races. Although he repeatedly insisted that his version of pre-Adamitism did not involve any polygenetic component—Adam was simply the descendant of pre-Adamite stock—it was widely believed, as the *Nashville American* reported, that Winchell's dismissal was on account of his adoption of "evolution and polygenism."[127] The hazardous brew of virulently anti-black sentiments, pre-Adamite thinking, and evolutionary theory to which Winchell gave voice so perturbed the Methodist confraternity that he was pushed out of Vanderbilt in an action that, like Woodrow's firing a decade or so later, made headline news and earned for him a place in the pantheon of scientific martyrdom. In the years that followed, any hint of pre-Adamitism or human evolution from a lower state, was likely to be seized upon by those zealous to guard the sacred flame of their own denominational tradition.

The Southern Presbyterian aversion to the idea of pre-Adamite races had even deeper historical roots. In his indictment of Josiah Nott in 1850, Howe had attacked as utterly deadly Nott's suggestion that only Caucasians were "the true Adamic race" and that "our servants are of another than the Adamic race, and of an inferior and brutish rather than human nature." That idea was wrong on every front. Howe's self-confessed "faith" in modern science was "too weak to receive the doctrine of Pre-Adamic races of men."[128] Besides, such talk not only contradicted Genesis; it undermined the true scriptural foundation of domestic slavery and thus the only authority worth having for the institution. Not surprisingly, the bulk of Howe's review was directed to a detailed exegetical defense of specific biblical passages against Nott's pernicious claim that the sacred writers were ignorant of geography, ethnology, natural history, and geology and that the Bible's chronology was suspect.

Despite their differences over biblical literalism, Woodrow and his devotees shared their opponent's racial biases. Woodrow himself declared in an 1888 piece on "Politics and Religion" that if his culture was "not to be blotted out of existence" in the aftermath of the Civil War, "the whites must rule."[129] He was no less sure that the Almighty had "divided the human race into distinct groups for the purpose of keeping them apart. And all attempts to restore the original unity of the race by the amalgamation of its severed parts have been signally rebuked."[130] Moreover, the Bible—"our sole and sufficient rule"—simply did "not condemn slaveholding." To be sure, he conceded, "slavery in the days of Christ and his apostles was not confined to any one race, and yet the rights of master and of slave are plainly taught in the New Testament." In this arena, he insisted, he had "no use for and little confidence in the ethnological argument."[131] Yet when he voiced his opinion on the slave system, he did not engage in detailed interrogation of scriptural passages. This omission is hardly surprising, of course, since he was on record as declaring that the Bible's task was to convey "moral, spiritual, and religious truth," not to teach science or history or geography.[132] It could hardly, in these circumstances, provide a blueprint for political philosophy. George Armstrong demurred, arguing that writers like Henry Drummond and J. S. Candlish who restricted revelation to matters of religious truth had given up the idea of plenary inspiration.[133]

It was similar with John Adger, one of Woodrow's chief defenders. As William Carrigan observes, "Religion played an important part in Adger's justification of the social order, but he did not rely upon a scriptural argument for support, rather opting for the theological argument that God had divinely inspired a race instinct."[134] Indeed, Adger was quite convinced that the annals of humankind had amply shown that God had "so constituted the two races as to make equality *forever* impossible."[135] He had long been persuaded—like John Bachman—that even though blacks were descendants of Adam they belonged "to an inferior variety of the human species." At the same time, he had "no sympathy with the new theory of a diversity of original races of men. We have no doubt whatever that the negro is of Adam's race."[136]

Neither Woodrow nor Adger, it seems, felt the need to ground their justification of southern slave culture in a literalist reading of the Mosaic narrative. For Dabney, Armstrong, Girardeau, and other opponents of evolution, it was different. When confronting the challenges that Darwinism posed, they lacked the wriggle room that a less literalist mindset made available. By the 1880s, Southern Presbyterians were still overwhelmingly wedded to a literal reading of the Bible as the sine qua non of their distinctive way of life. Any tinkering with the scriptural foundations on which their culture rested ran the risk that the whole edifice, already creaking under the strains of defeat in the Civil War, would come crashing down around them.

♦ ♦ ♦

OCCUPYING A PROMINENT POSITION in the culture of the Old South, antebellum Presbyterians had come to regard biblical orthodoxy as the foundation stone of the southern social order. During the 1850s this happy coalition came increasingly under threat with the rise of antislavery sentiments. A crucial component in the ideological apologetics of southern Presbyterians was their conviction that an honest-to-goodness, unadulterated reading of the Bible provided ample warrant for the institution of slavery and, later, for racial segregation. Abolitionist attacks on the South were seen as rationalistic assaults on the integrity of scripture and the Christian character of southern culture. The Bible was thus appealed to as a means of resisting a host of perceived Yankee evils—radical democracy, emancipation, higher criticism, and modern science. These were seen as subversive of what was taken to be a biblically sanctioned southern culture and as promoting godless notions of human equality. In these circumstances, any tampering with the Mosaic chronicles of creation was culturally and politically explosive, not least when the proposal to engage in fancy hermeneutic footwork was to make peace with the latest whim emanating from secularized science.

Relinquishing a literal account of Adam's creation and the origin of humankind, moreover, had already done irreparable damage. It had bestialized African Americans by defining an entirely separate lineage for the race, and in some extreme cases it even ruled them outside the arc of humanity itself. Southern Presbyterians had no use for any such dogma, even if it conspired to establish black racial inferiority. They had already, in the antebellum period, worked out what they considered to be a persuasive scriptural justification for slaveholding and convinced themselves that it was benign paternalism. Scripture consecrated slavery; secular science corrupted it. In the postbellum period, Southern Presbyterian apologists for the Lost Cause of the Confederacy saw no reason to renege on their traditional stance on human genesis. Having defended the teachings of scripture against the perverse anthropology of Nott, Gliddon, Morton, and the rest of that clergy-baiting crowd, they weren't going to sacrifice its literal meaning to the likes of Darwin, Tyndall, Huxley, and Spencer. Everything they had seen confirmed to the vast majority, not least those older churchgoers whose memories of the Old South still burned bright, that Darwinian evolution was simply the newest garb under which treacherous science masqueraded in its ongoing plot to demolish the foundations of southern civilization. This is the context in which the Woodrow affair at the Columbia Seminary was domesticated. It was exceptionally difficult for interlocutors—especially those intellectual leaders of a former generation—to acquiesce to the kind of rereading of scripture that adopting evolution required. It was even harder to do so in a situation where cultural identity was so tightly moored to biblical literalism and where for decades already southern Presbyterians had fought to retain the authority of Genesis in the face of the scurrilous sniping of an overweening anthropology.

CHAPTER SIX

Princeton, Darwinism, and the Shorthorn Cattle

FOR A MONTH OR TWO in 1873, just before he entered Princeton Theological Seminary in September to train for the ministry, Benjamin Breckinridge Warfield (1851–1921) acted as livestock editor for the *Farmer's Home
Journal* of Lexington, Kentucky.[1] In many ways, Warfield, who would later assume the chair of didactic and polemic theology at Princeton Seminary, was an
ideal choice for the position. In 1868 he had entered the College of New Jersey
(later Princeton University), where he studied mathematics and scientific subjects
before graduating with highest honors in 1871. And if that did not in itself fully
equip him for the job, his personal predilections certainly helped. His brother
later recalled how avidly he "collected birds' eggs, butterflies and moths, and geological specimens" and the enthusiasm with which he "studied the fauna and flora
of his neighborhood; read Darwin's newly published books with enthusiasm; and
counted Audubon's works on American birds and mammals his chief treasure."[2]

Warfield maintained his interest in natural history, and more particularly in
livestock breeding, for the remainder of his life. Over many years, he carefully
kept a substantial scrapbook of clippings—mostly of his father's writings—on
what he called "Short Horn Culture."[3] His father, William Warfield, was a leading
farmer of old Whig stock in the Kentucky bluegrass region and continued the
family interest in the breeding of shorthorn cattle.[4] The family's herd, it was said,
had a longer and more interesting history than any other in America: "The record
of its victories in the show ring would fill a small volume."[5] Already well known
through the sales ring and the livestock show, William acquired a vastly wider
reputation through such writings as *American Short-Horn Importations* (1884)
and *The Theory and Practice of Cattle-Breeding* (1889).[6] These publishing ventures
were a family undertaking. In the preface to the 1889 book, William paid tribute
to "the assistance my sons have given me in preparing all my work for the press.
Without their aid much—even most—of it could never have been done." "Great
credit is also due to my elder son, Prof. Benjamin B. Warfield, D.D., now of Princeton, N.J.," he went on, "whose energy and vigor of thought and pen gave me such
essential aid in the earlier years of my connection with the press; nor has the

pursuit of the more weighty things of theology destroyed his capacity for taking an occasional part in the active discussion of cattle matters. The papers which have appeared over my signature have thus to quite a large degree been of family origin."[7]

Judging by *The Theory and Practice of Cattle-Breeding*, the Warfield family was well acquainted with the contributions of Charles Darwin. This is scarcely surprising, of course. Matters of variation, heredity, reversion to type, and such like, occupied the thoughts of animal breeders—whether of pigeons or pigs, canaries or cattle—as much as they aroused the curiosity of Darwin, not least in his *The Variation of Animals and Plants under Domestication*. Indeed, Darwin's thoughts on all sorts of subjects are laced into the fabric of Warfield's book. Far more than any other authority cited, Darwin's name surfaces over twenty times on such matters as line breeding, heritable mutilations, the value of crosses, prepotency, atavism, and variation.

It was not just empirical particulars that William culled from Darwin's writings however. Natural selection was a fundamental principle:

> Nature's method seems to be a wide and general system of selection, in which the strong and vigorous are the winners and the weaker are crushed out. Among wild cattle the more lusty bulls have their choice of the cows in a way that under natural selection insures the best results to the race . . . If the conditions of life should suddenly change the result on such wild cattle would be to deteriorate or to improve the average according as the change was for their advantage or disadvantage. It is quite apparent that no question of breeding intrudes itself here. Nature's selection, while always in favor of the maintenance of the animals in the best manner, yet is impartial, and under ordinary circumstances would maintain an average.[8]

This affirmation reinforced his earlier depiction of the way in which the "economy of Nature" operated. "By 'natural selection' the strongest are made stronger; the weaker go to the wall. The survival of the fittest was a well chosen and apt term to express this idea. On some soils one plant will thrive and displace others which would displace it in a different soil. In one climate one variety of animals finds a congenial home while others pine and die."[9] Like Darwin himself, then, Benjamin Warfield encountered natural selection through an intense study of breeding.[10] His father, William, himself said as much. "Studies of the laws of heredity, of natural selection, and many other specific problems have won years of devoted labor from many active scholars. What the scientist has approached from the side of theory the practical breeder has assailed on the side of practical every-day utility. The studies of the one have borne their fruit in the application of their results to the labor of the other."[11] This, I believe, is why Benjamin could describe himself, during his student days, as a pure Darwinian—he first met the theory empirically, as it were, not theologically or philosophically.

Just as Benjamin Warfield enrolled as a student at Princeton in 1868, James McCosh (1811–1894) arrived from Belfast to take up the presidency of the College of New Jersey. Reflecting on Dr. McCosh's arrival in Princeton, Warfield observed: "No, he did not make me a Darwinian, as it was his pride to believe he ordinarily made his pupils. But that was doubtless because I was already a Darwinian of the purest water before I came into his hands, and knew my *Origin of Species* and *Animals and Plants under Domestication*, almost from A to Izard."[12] With cattle breeding in his blood, he could hardly have missed the relevance of Darwin's researches for a family business in which his father kept abreast of the latest scientific thinking on the whole subject. To be sure, later in life, Warfield told his readers, he rather departed from McCosh's "orthodoxy" on Darwinian matters, but there is good reason to suppose that this was because in the decades around 1900 Darwinism found itself in eclipse alongside competing evolutionary narratives. As I have pointed out, neo-Lamarckism, orthogenesis, the mutation theory of Hugo de Vries, and other evolutionary schemes were in current circulation, but they did not assume the gradualism or the central significance of natural selection which Darwin's disciples so energetically promulgated.[13] In 1868, however, Warfield had no doubts. That year, on 1 May, he purchased his own copy of Darwin's *The Variation of Animals and Plants under Domestication*, and within a couple of years he would acquire *The Origin of Species*, *The Descent of Man*, and *The Expression of the Emotions*.[14] Besides, for Warfield, McCosh was an "inspiring personality" and a "great teacher" . . . even if he "completely lacked a sense of humor." He was, Warfield confessed, "the most inspiring force that came into my life during my college days."[15]

But Warfield was subject to another charismatic Princeton presence as well. When he returned to New Jersey in 1873, this time to the seminary, he came under the magnetic power of the aging Charles Hodge (1797–1878). Warfield quickly adopted Hodge as his theological mentor, and later, in 1887, when he took up a chair at the seminary, he continued to use Hodge's *Systematic Theology* as his staple teaching text. During the winter of Warfield's first term at the seminary in 1873, Hodge was busily writing what would turn out to be his last book. That October, he had spoken out against Darwinism at the New York meeting of the Evangelical Alliance, and over the following months he turned his preliminary reflections into a 178-page volume entitled *What is Darwinism?* The book delivered Hodge's answer with crystal clarity: "It is atheism."[16]

On the surface at least, that judgment could not have been further from James McCosh's. So as Warfield contemplated the verdicts of these two giants on the Princeton intellectual landscape, he witnessed two entirely different strategies for engaging with the Darwin phenomenon. Just how different the essence of their views was is certainly open to question.[17] Both, for example, were determined to resist any removal of design from the natural order; teleology, not transmutation, was the key metaphysical issue. But the *rhetorical* stances they adopted—the poetics of their pronouncements—could not have been more different. For it was style of communication, rather than substance of evaluation, that secured for one a reputation as the era's foremost reconciler of Christianity and Darwin and staged the other as the most astute theological anti-Darwinian of his time.

Rivals in Rhetoric

By the time McCosh arrived in New Jersey, he was a leading moral philosopher in the Scottish mode and of considerable standing. He had come to Princeton directly from Queen's College, Belfast, where he had held the chair of logic and metaphysics since 1851. Prior to that he had been a Free Church of Scotland minister in Brechin and had thrown himself vigorously into the reforming movement that came to a head in the Disruption of the established church in 1843. His crossing the Irish Sea to take up the Belfast chair was the consequence of good fortune and forceful scholarship. Good fortune manifested itself in the fact that Robert Blakey, the first occupant of the Queen's chair, seemed to prefer fish to philosophy. Reportedly dismissed from his position for persistent failure to show up to his classes, Blakely found time to author four books on angling, including several guides to fishing excursions in England, Scotland, Wales, France, and Belgium.[18] To be fair, though, he also published a range of philosophical works,[19] which received the acclaim of the prolific French thinker Victor Cousin (1792–1867) in his account of *La Philosophie Écossaise*.[20] Still, Blakey's departure opened the door to McCosh's academic career.

As for forceful scholarship, the intellectual stature of McCosh's first book, *The Method of the Divine Government* (1850) impressed itself so strongly on the mind of George Villiers—fourth Earl of Clarendon and Lord Lieutenant of Ireland—that one Sunday morning, while reading it, he forgot to go to church. Blameworthy though this may have been in the eyes of Belfast Presbyterians, Clarendon doubtless made up for it by offering McCosh the vacant Queen's chair. In this work, McCosh scrutinized John Stuart Mill's *System of Logic* with critical eyes and urged that morality had to do with internal motivation, not with external actions. Any ethical theory that applied measurements to mere results was wrong-headed.

Moral integrity was not a matter of calculus. It was just mistaken to think a person virtuous simply because they happened to do the right thing. McCosh was sure that polite society's preoccupations with decorum and civilized conversation should not be confused with true virtue. All these beliefs disclosed his long-standing enthusiasm for Scottish Common Sense philosophy. Determined as he was to "rip the mask off the face of the moderate age" and expose its self-satisfactions,[21] he followed in the footsteps of Thomas Reid, who, when analyzing what he called "the intellectual powers of man," pointedly declared: "On the one side, stand all the vulgar, who are unpractised in philosophical researches, and guided by the uncorrupted instincts of nature. On the other side, stand all the philosophers ancient and modern . . . In this division, to my great humiliation, I find myself classed with the vulgar."[22] Reid intended this declaration to mark his impatience with the tortured meanderings of so-called geniuses whose speculations led to sceptical absurdities. At the same time, it gave voice to his enthusiasm for the epistemic principles taken for granted in the common affairs of everyday life.[23] McCosh concurred. When it came to choosing between the speculative abstractions of the high and mighty and the honest-to-goodness common sense of every Tom, Dick, and Harry, McCosh knew what side he was on.

That enthusiasm for inductive particulars governed McCosh's thinking on the sciences of mind and nature. Only by the rigorous prosecution of induction could any intellectual progress be made. As we saw in chapter 3, his passion for Baconianism conspicuously manifested itself during his inaugural lecture in Belfast in 1852. With these sentiments, it is not surprising that McCosh would shortly join forces with the Queen's College professor of natural history, fellow Scotsman, and specialist in the study of algae, George Dickie, to produce *Typical Forms and Special Ends in Creation* (1855). By that stage, McCosh had already been engaged in independent investigations into plant morphology and had spoken on the subject at several meetings of the British Association.[24] In some ways the McCosh-Dickie volume can be read as the high watermark *and* the last gasp of empirical natural theology before the challenge of Darwinism. But this would be to underestimate its significance. The book was largely patterned on the perspective of the comparative anatomist Richard Owen, who believed that organic structures were organized according to transcendental plans and that corresponding structural forms (homologous relations, as they were called) were to be found among different animal families.[25] This idealist system,[26] especially when used to supplement standard Paleyite adaptationism, was immensely suited to the McCosh-Dickie project, providing as it did resistance to the onslaught of the Paris materialists.[27] It also enabled McCosh in the long run to negotiate a much more positive response to Darwin than many others who restricted themselves to the more conventional forms of teleology.[28] For the purpose of the whole book was to show how, of late, "writers on the theology of nature" had been "unnecessarily narrowing" the argument from design. In so doing, they compared unfavorably with the "sublime philosophy of Plato," who had shown how design could be detected in the way in which "all things are formed according to unalterable laws or types." Returning to design by natural law was fundamental to the McCosh-Dickie venture. As they put it, "All things in this world are subordinated to law, and this law is just the order established in nature by Him who made nature, and is an order in respect of such qualities as NUMBER, TIME, COLOUR, and FORM." To them this represented "a class of phenomena to which Paley has never once alluded in his Natural Theology" and revealed the brilliance of Owen, who in their own day had succeeded in developing "a teleology of a higher order than Cuvier."[29]

By locating design every bit as much in archetypal plans and lawlike regularity as in the specific adaptations of the sort that animated Darwin's investigations, McCosh came to see that the causal operations of natural selection as a law of nature could be allowed free play. His thinking on the nature of "chance" was also important as he sought for a via media between advocates of radical contingency and champions of rigid determinism. To McCosh and Dickie there certainly was "such a thing as chance," but at the same time there were "adjustments in nature which can not proceed from chance."[30] In this interplay between the random and the determined, McCosh found space to think of natural selection as a mechanism that held together the lawlike and the contingent. This work with Dickie in Belfast laid the ground for McCosh's later emergence as perhaps the foremost reconciler of evolution and Protestant theology.[31]

It was doubtless the quality of works like this, as well as *Intuitions of the Mind Inductively Investigated* (1860), *The Supernatural in Relation to the Natural* (1862), and *An Examination of Mr J. S. Mill's Philosophy* (1866), that brought McCosh to the attention of the trustees of the College of New Jersey as they sought for a successor to president John Maclean. During the inauguration ceremonies on 27 October 1868, Marcus Ward, governor of New Jersey and ex officio president of the board of trustees, reviewed McCosh's publications and discerned in them the "great depth of thought and the erudition of a mighty scholar."[32]

Within a few years of arriving in the New World, McCosh was invited to deliver the Ely lectures at Union Theological Seminary in New York City. Taking as his overall theme "Christianity and Positivism," he turned in his second and third lectures to the general subject of evolutionary development. Here he announced that while he was doubtful about its universality, he was certainly inclined to think that the theory of natural selection "contains a large body of important truths, which we see illustrated in every department of organic nature." Natural selection was a principle clearly "exhibited in nature, and working to the advancement of the plants and animals from age to age."[33]

But there were definite limits to its power. In the field of ethics, for example, McCosh strenuously resisted any resort to Darwinian explanations. The vulnerable in society needed protecting, not exterminating. "The law of the weak being made to give way before the strong" he observed, "is very apt to be abused and will certainly be perverted by those who do not take into account the other and higher laws which limit it, and are expected to subordinate it. If they look to it alone, they will understand it as meaning that the poor and the helpless need not be protected or defended, but may be allowed to perish." To McCosh the political consequences were entirely pernicious. The idea that "there are essentially inferior races, which are doomed to give way 'in the struggle for existence'" was attracting "defenders" who looked "on the prospect with complacency, provided a few favored races are enabled to advance on 'the principle of natural selection.'" Such sentiments, he insisted, were currently "exercising, directly or indirectly, a very injurious influence on public sentiment in this country and in others."[34] McCosh's strongly abolitionist sympathies made him reject scientific schemes that might underwrite racialist ideologies.

In an appendix on *The Descent of Man*, which had appeared while McCosh's lectures were going through the press, he further expanded on his feelings about human evolution. To begin with, he acknowledged "a striking analogy between man and the lower animals, between all the tribes of animals, and between animals and plants" and noted Darwin's impressive contributions to the subject. But he remained chary of the idea that "the animal is evolved from the plant, and man from the lower animals." However powerful an explanatory mechanism natural selection was, McCosh was sure that it could not explain "how Life arises, or Sensation, or Consciousness, or Intelligence, or Moral Discernment." Scientifically, he found Darwin's theory of gemmules as the vehicle of inheritance much too vague. Psychologically, he found his theory of human descent from primates much too disturbing: "I confess I shrink from it," he conceded.[35]

Charles Hodge shrank from it as well—though more comprehensively. In 1862 Hodge had given his first word on Darwinism in an article dealing with recent anthropological attacks on the unity of the human species by such scientists as Samuel George Morton, Josiah Nott, George Gliddon, and Louis Agassiz.[36] Toward the end of the piece, after a lengthy rebuttal of Nott and Gliddon's *Types of Mankind*, Hodge delivered a few remarks on an equally troubling attack from "the opposite pole of skeptical speculation in natural history"—Darwin's *Origin of Species*. "The object of this interesting work," Hodge observed, "is to prove that there is no such thing as permanence in the species of natural history." Thus while the polygenists "find such differences between man and man, that the different groups could never have descended from a single pair, . . . Darwin finds so little difference between man and the animals, that he believes them all to be 'descended from at most only four or five progenitors,' and infers, 'from analogy,' that they are all 'descended from some one primordial form.' "[37] Plainly, Hodge immediately cast natural selection as a subversive thesis about human origins and, regardless of its monogenetic implications, located it in the heterodox neighborhood of southern polygeny.

Ten years later, in 1872, the first two volumes of his monumental *Systematic Theology* appeared. By now Hodge enjoyed an unrivaled reputation as the leading exponent of historic Calvinism in the United States and the foremost advocate of Old School confessional Presbyterianism. In the first volume, after an expansive methodological prolegomenon, he turned to evaluate what he labeled "anti-theistic theories," among which the scientific materialism of Huxley and Spencer loomed large. Darwin's judgment had to wait for the second volume, when Hodge took up the subject of human origins. In less than one page he dealt with the biblical doctrine. Nearly thirty pages rebutting what he called "Anti-Scriptural Theories" followed. Here he reviewed a range of subjects, including ideas about spontaneous generation and several pre-Darwinian evolutionary narratives, before directing attention specifically to Darwin's theory. Right away, Hodge noted that Darwin stood "in the first rank of naturalists, and is on all sides respected not only for his knowledge and his skill in observation and description, but for his frankness and fairness." Next, he gave an informed summary of the principles—variation, heredity, natural selection, and the struggle for life—that lay at the heart of Darwin's theory. Hodge found the "system" to be "thoroughly atheistic." This, he hastened to add, did not mean "that Mr. Darwin is an atheist." Nor indeed that "every one who adopts the theory does it in an atheistic sense." It was perfectly conceivable that there might "be a theistic interpretation of the Darwinian theory." But theistic evolution was not Darwin's own theory, and the efforts of Asa Gray, the Harvard botanist, "to vindicate Darwin's theory from the charge of atheism" were unconvincing and misguided. Darwin's theory was inherently anti-teleological; natural causes were undirected; development took place without "reference to an end." In Darwin's universe, "Teleology, and therefore, mind, or God, is expressly banished from the world."[38]

For all McCosh's reservations about Darwinian imperialism, by contrast, when he came to address the sixth general conference of the Evangelical Alliance, which convened in New York in 1873, his rhetoric was anything but antagonistic. He just sounded very different from Hodge. His was the first presentation in the Philosophic Section of the congress on Monday, October 6, a day devoted to "Christianity and its Antagonisms." Questions of Darwinian evolution dominated the session. McCosh began by reviewing the biblical portrayal of cosmogony as laid out in the Genesis narrative and then juxtaposed this picture with modern science. "A somewhat different but not inconsistent view is given of the same objects on the scientific side," he reported, "where every thing is ascribed to what is called Law, which, however, when properly understood, implies a lawgiver. So these men, consciously or unconsciously, are unfolding to our view the plan of the great Creator." As he pursued this line of thinking, McCosh turned to the debate between those who maintained that the appearance of new species required direct divine action, and those holding that there were powers in nature "which gradually raise species into higher forms by aggregation and selection." McCosh doubted that religion had any "interest in holding by the one side or the other of this question, which it is for scientific men to settle." Nor indeed was religion "entitled to insist that every species of insect has been created by a special fiat of God, with no secondary agent employed." Of course there were limits to evolution's empire. But McCosh was concerned about the spiritual consequences of outlawing evolution tout court, as too many were now doing: "It is useless to tell the younger naturalists that there is no truth in the doctrine of development, for they know that there is truth, which is not to be set aside by denunciation. Religious philosophers might be more profitably employed in showing them the religious aspects of the doctrine of development; and some would be grateful to any who would help them to keep their old faith in God and the Bible with their new faith in science."[39]

Hodge also had shown up to the New York gathering to address the Alliance on the subject of "The Unity of the Church."[40] But he also found himself in McCosh's session and, after the Scotsman's presentation, intervened to ask one of the other commentators, Rev. J. C. Brown, what he considered to be a fundamental question—one that separated "theists from atheists—Christians from unbelievers." Was "development an intellectual process guided by God," or was it "a blind process of unintelligible, unconscious force, which knows no end and adopts no means?" This was the "vital question." "We can not stand here and hear men talk about development," he went on, "without telling us what development is." To Hodge, matters of definition were fundamental to assessing the meaning of such terms as *evolution, natural selection,* and, most particularly, *Darwinism.* But his question was not intended to elicit a response that would inform his own thinking; he was already sure of the correct answer. "My idea of Darwinism," he observed a little later in the discussion, "is that it teaches that all the forms of vegetable and animal life, including man and all the organs of the human body, are the result of unintelligent, undesignating forces . . . Now, according to my idea, that is a denial of what the Bible teaches, of what reason teaches, and of what the conscience of any human being teaches . . . It excludes God; it excludes intelligence from every thing."[41]

In the wake of this skirmish, Hodge settled down to compose a book-length evaluation of Darwin's theory that appeared the following May, in 1874. The pace of *What is Darwinism?* was more leisurely, the empirical detail more expansive, the argument more extended, but the verdict was identical—atheism. The whole project was an extended exercise in definition. To Hodge the *nature* of Darwinism lay in its *definition*. Because he was certain that Darwin's use of the word "natural" was "antithetical to supernatural," Hodge insisted that "in using the expression Natural Selection, Mr. Darwin intends to exclude design, or final causes." Here the essence of the theory lay exposed. That "this natural selection is without design, being conducted by unintelligent physical causes," Hodge explained, was "by far the most important and only distinctive element of his theory." In a nutshell, the denial of design was the very "life and soul of his system" and the single feature that brought "it into conflict not only with Christianity, but with the fundamental principles of natural religion."[42] By this definitional move, Hodge could set the terms of the debate. To control definitions, of course, is to exercise power. In Hodge's case, it meant that he could adjudicate on who was and was not a Darwinian. Those like Asa Gray who considered themselves Christian Darwinians were either mistaken or just plain mixed up; that label had no meaning. Thus for all his efforts to teleologize Darwinism, Gray simply was "not a Darwinian."[43] That he was a Christian *evolutionist,* Hodge had no doubt, but that was an entirely different matter. Darwinism *was* atheism.

The following year McCosh reproduced and substantially expanded his Evangelical Alliance talk, this time under the simple title "On Evolution." Printed as an appendix to Rev. J. G. Wood's *Bible Animals,* the tone of McCosh's verdict on Darwinism remained unaltered in the wake of Hodge's pessimistic diagnosis. For a start, while the Evangelical Alliance presentation never once referred to Darwin*ism,* his first couple of sentences in Wood's volume brought together in no uncertain terms development, evolution, and Darwinism. "In these days" he began, "every educated man and woman talks of development, of evolution and of Darwinism. Many are anxious to know what they are, whether they are established by scientific evidence, and what is their moral and religious tendency. In this paper, without entering into minute scientific details, I am to give a plain account of this new theory." Besides, while he considered that "Darwinism is . . . encompassed with many difficulties, and cannot be regarded as established," he was no less certain that "there may be truth in it, and yet it may require to be greatly modified." McCosh had no doubt that "evolution runs through all nature: one thing comes out of another." But this did not mean that Darwin's theory reached beyond the sphere of the hypothetical. For one thing, species transmuta-tion was not empirically observable. Nor did natural selection enjoy anything like the universal scope that partisans frequently claimed; it could, for example, "offer no explanation of the origin of the matter out of which animated beings are formed." Siding with A. R. Wallace, he insisted that there were immeasurably greater empirical obstacles in the way of Darwin's theory of human evolution. McCosh's overall judgment was that "the Darwinian theory as a whole is not proven, and that it will need to be greatly modified, limited and enlarged before it is entitled to command our assent."[44]

McCosh's verdict certainly did not sound like a ringing endorsement, but nei-ther did it have the disapproving tone of Hodge's ruling. The rhetoric was alto-gether different. Because scientific inquiry was captivating an increasing number of younger people, it was vital that they should be allowed scope to "accept the truths of science, so far as they are established" yet without being "captivated by theories which go far beyond the facts, and which may require to be modified and corrected." To be sure, there would be "times when there seems to be a contradic-tion between science and religion" and frequently occasions when "we cannot see the reconciling link." In such circumstances, patience was needed. Rather than jumping to rash conclusions or issuing edicts that could cause irreparable damage, spokesmen should "wait for further light" when any "seeming incon-gruity between Genesis and geology" came to the fore.[45]

During the autumn of 1874, McCosh had returned to Princeton from a visit to Britain. Just before leaving, he had got hold of a copy of Tyndall's infamous Belfast Address and read it on the journey home. His response became the first lecture in his course on the history of philosophy that fall, and the written version found its way into the *International Review*. The New York publisher Robert Carter also printed it as a booklet in 1875. McCosh was quick to point out that he agreed with Tyndall's conviction that "much good may be done in Ireland by the spread of scientific knowledge, as fitted to lessen the bitterness of ecclesiastical feuds." But he was sure that Tyndall's tactics had only succeeded in adding "a new ele-ment of trouble into the boiling caldron, and . . . [had] thrown back the general study of physics in Ireland." A good deal of McCosh's analysis centered on Tyndall's history of atomic theory, and he did not hesitate to identify what he considered to be those "blunders" that "crop out ever and anon."[46] He also registered profound concerns, much like Montesquieu, over the social consequences of Epicurean philosophy: the erosion of the mind-matter dualism; the collapsing of Aristotle's four causes; the evacuating of teleology from biological explanation; and the effort to reduce sensation, judgment, reason, love, passion, and resolution to atoms and molecules. Surely Tyndall himself couldn't think that the thoughts and theories that abounded in his own Belfast address were nothing but matter in motion.

Yet these metaphysical accretions did nothing to diminish McCosh's belief that the "doctrine of development" was one of the two "great scientific truths" of the century—the other being the conservation of energy—and that it had triumphed to "an extent which was not dreamed of till the researches of Darwin were published." Even in the wake of Hodge's verdict, Darwin's anti-teleological bias did not erode McCosh's confidence in the reality of evolution. For he was certain that it "could be easily shown that the doctrine of development, properly understood, and kept within inductive limits, is not inconsistent with final cause." To McCosh, a "determined order" was easily detectable, and in this he drew sustenance from Leibniz's ideas of preestablished harmony. "In due time," he was sure, "a Paley will arise to furnish proofs of design" in the new Darwinian universe. "Darwin will supply the facts" he went on, "and we are just as capable as he of perceiving their meaning. He may reject teleology, but his facts are teleological whether he acknowledges it or no."[47]

Rhetorically speaking, McCosh's depiction was a far cry from Hodge's "Darwinism is atheism" declaration. There is no doubt that the title of his 1887 Beddell Lectures, *The Religious Aspect of Evolution*, was one that Hodge would never have chosen. Right from the start, McCosh made it clear that he had always held to the view that "Darwin was a most careful observer, that he published many important facts, that there was great truth in the Theory, and that there was nothing atheistic in it if properly understood." Now, many years on, he was still concerned to keep abreast of scientific developments and had run the manuscript of his lectures past George Macloskie and William Berryman Scott (Charles Hodge's grandson), professors, respectively, of natural history and geology at Princeton. Like them, he read evolutionary history through the lens of ultimate purpose. As he put it: "The design is seen in the mechanism. Chance is obliged to vanish because we see contrivance. There is purpose when we see a beneficent end accomplished. Supernatural design produces natural selection." To McCosh, there was "nothing atheistic in the creed that God proceeds by instruments." The mechanisms of change that the Creator had installed in nature showed his purpose every bit as clearly as any miraculous intervention. This understanding meant that the "principle of the survival of the fittest is a beneficent provision" and, consequently, that the "religious man should not object to it, if at certain junctures it produces a newer and higher species of plant or animal to make up, it may be, for the disappearance of an old species." For himself, McCosh confessed that he had "never been able to see that religion, and in particular that Scripture in which our religion is embodied, is concerned with the question of the absolute immutability of species. Final Cause, which is a doctrine of natural religion, should be satisfied with species being so fixed as to secure the stability of nature."[48]

The misunderstanding of secondary causes by religious believers, of course, was no new phenomenon, as McCosh well knew. "It is a forgotten circumstance," he remarked, "that when Newton proclaimed the law of gravitation it was urged that he thereby took from God an important part of his works to hand it over to material mechanism, and the objection had to be removed in a quarto volume written by the celebrated mathematician, Maclaurin; and this was the more easily done from the circumstance that Newton was a man of profound religious convictions. The time has now come when people must judge of a supposed scientific theory, not from the faith or unbelief of the discoverer, but from the evidence in its behalf." Now the scientific arena had changed, as had the key players on the stage. But the same confusion prevailed. "Because God executes his purposes by agents, which it should be observed he has himself appointed," he explained, "we are not therefore to argue that he does not continue to act, that he does not now act. He may have set agoing the evolution millions of years ago, but he did not then cease from his operation, and sit aloof and apart to see the machine moving. He is still in his works, which not only were created by him, but have no power without his indwelling."[49]

McCosh was fully aware of the difficulties in prosecuting a case like this in a persuasive way, and he carefully circumscribed his reconciling tactics. In answer to the question "Is the Method of Evolution a Good One?" he made it clear that he was "not prepared to prove that evolution is the best way in which God could have proceeded, or that there are no other ways equally good in which he acts in other worlds." "All that I profess to do," he insisted, "is to show that the method is not unworthy of God; that it is suited to man's nature; that it accomplishes some good ends. It is to this extent that I would 'justify the ways of God to man'." This was, at best, a tentative evolutionary theodicy. For McCosh well knew that "in looking on nature as the work of God we meet with perplexities." Unanswerable questions abounded. "How are certain evils, disease and death, and inevitable sorrow consistent with the justice and love of God?" he pondered. He fully admitted "that there are results following from the laws of God, which it is not easy to reconcile with the omniscience and benevolence of Deity."[50]

Whatever its scientific merits, however, and McCosh was sure that it had many, it was evolution's *spiritual* potential that ultimately galvanized his 1887 Bedell Lectures at Kenyon College, Ohio. In ways somewhat akin to Henry Drummond's project of spiritualizing evolution,[51] McCosh wanted to locate evolutionary progress in the wider framework of eschatological destiny. As he peered into a dimly lit future, he found in evolution a clue to the coming age. "In all the geological ages we find in any age the anticipation of the following," he mused. "This may also be the case with the age in which we now live, the Age of Man. We see everywhere preparations made for further progress, seeds sown which have not yet sprung up; embryos not yet developed; life which has not yet grown to maturity. In particular we find that in this Age of Man, man has not yet completed his work."[52] For McCosh, the events of the *Heilsgeschichte* of the Christian era were to be located in the wider context of the progressive development of the great chain of life. The advent of the human species and the coming of the Spirit at Pentecost were to be understood as inaugurating new stages of human existence in what he called "the dispensation of the spirit." As he explained, "In all past ages there have been new powers added. Life seized the mineral mass, and formed the plant; sensation imparted to the plant made the animal; instinct has preserved the life and elevated it; intelligence has turned the animal into man; morality has raised the intelligence to love and law." This meant that the "work of the Spirit is not an anomaly. It is one of a series; the last and the highest. It is the grandest of all powers."[53] The drama of what he called "the Age of Man," namely, the epochs through which the human race had passed, were thus nothing less than the unfolding of spiritual evolution. Whatever the drawbacks, Darwinian and Lamarckian natural history presented McCosh with a rich set of concepts from which he constructed an eschatological hermeneutic that made sense of humankind's spiritual journey from the brutal struggles of primordial "man" to the glories of the coming eschaton.

At the fundamental level, James McCosh was every bit as concerned about Darwinian naturalism as Charles Hodge. In significant ways, they saw the world the same way. So it was no surprise when, during the inauguration ceremonies of McCosh as Princeton president in 1868, Charles Hodge, the most senior member of the board of trustees, presented the formal welcome in the warmest of tones. "In no case within our knowledge," he began, "has an academic election been received with such unmistakable evidence of public approbation." This judgment, he reported, was based on "the high positions" McCosh had already successfully filled, the "world-wide reputation" he had acquired for scholarship, and "personal knowledge" of his upright character.[54] Four years later, in 1872, when a special commemoration was held to mark Hodge's fifty-year occupancy of his chair at the seminary, McCosh was able to reciprocate. Appointed to speak for three branches of "that old Church of Scotland who is the mother of us all," he was particularly pleased to report that his own Free Church of Scotland "looks on Charles Hodge and her own William Cunningham, as the greatest theologians of this age."[55]

Despite the mutual admiration, the rhetorical flavor of their pronouncements on evolution could not have been more different. Nor their legacies. Hodge's reputation as a resolute anti-Darwinian was well established. When Josiah L. Porter from Belfast rose to speak during the celebrations of Hodge's half-century professorship, he made a point of emphasizing that "Dr. Hodge has taught us . . . how we are to show to Darwin, and Colenso, and others of that school, that we have a philosophy better than theirs, that we have principles nobler than theirs, and that we are able to meet them on their own ground." In Porter's circle, Hodge's name was without peer. And so he was keen to recruit him for other campaigns, too. Given what he called the "special circumstances" prevailing in Ireland, he was determined to enlist Hodge in the crusade "to smite to the earth the colossal fabric of Popish tyranny in poor, oppressed Ireland!" For, as he explained, the "two gigantic systems of error" against which Ulster had to contend were "Popery on the one hand . . . Infidelity on the other." And "in the battle which we have to wage with each of those," he declared, "Dr. Hodge has rendered to us, as he has rendered to you, the most signal service."[56]

By contrast, because he expended considerable intellectual energy in developing an evolutionary teleology that conceived of organic progress along predetermined paths, McCosh secured for himself a different legacy. Whatever reservations he entertained about Darwinian naturalism, the rhetoric he employed meant that he was routinely cast as perhaps the foremost Christian Darwinian of his age. The *New York Times,* for example, told its readers on 20 April 1890 that "Dr McCosh is one of the few clergymen . . . who have been willing to accept Charles Darwin's theory of evolution from the very first."[57] Rev. George Macloskie, a former student and later professor of biology at Princeton, recalled that McCosh's insights into "the religious bearing of Darwinism" and "his acceptance of it when properly understood" had done much to avert a potentially "disastrous war between science and faith." That, Macloskie contended, meant that "in 'his' college, men have studied Biology without discarding their religion." Others concurred. The distinguished geologist, explorer, and Berkeley professor Joseph Le Conte, wrote to McCosh in 1888. "I am convinced that you are doing a good and very important work," he said, "in showing that evolution is not necessarily atheistic, nor in any way antagonistic to a true religious belief." None other than Andrew Dickson White concurred. "In one of his personal confidences," he wrote of McCosh, "he has let us into the secret of this matter. . . . [H]e saw that the most dangerous thing which could be done to Christianity at Princeton was to reiterate in the university pulpit, week after week, declarations that if evolution by natural selection, or, indeed, evolution at all, be true, the Scriptures are false. McCosh tells us that he saw that this was the certain way to make the students unbelievers."[58]

This issue was not the only one in which style of communication rather than substance of argument shaped history's judgment of McCosh. He was a past master at using words wisely. Back in Ireland, during the momentous events surrounding the outburst of religious fervor in 1859, he had found it possible to query the Ulster revival's "physiological accidents" and yet be seen as the key theological champion for the awakening. Thus he insisted that the profound mental turmoil that converts exhibited—and that manifested itself in dramatic bodily seizures—was of divine origin. These convulsions were simply accidental accompaniments to genuine religious experience, even if they expressed emotional excess and psychosomatic disturbance.[59] Here, too, McCosh's scientific investigation sounded like theological vindication. Whether speaking of divine visitation or Darwinian vision, rhetorical nuance counted as much, if not more, than cognitive content.

McCosh knew only too well that the craft of careful wordsmithing was critical to the career of a university president. Embarked as he was on the task of modernizing the college through improving the science curriculum without abandoning its traditional heritage, he needed to balance the forces of tradition and progress with sensitivity. Even if he was perhaps the most distinguished scholar to fill the post since Jonathan Edwards, he knew that some of the more conservative college trustees had their doubts about him.[60] In these circumstances, the need to steer a prudent course was nowhere more vital than on the evolution question. George Marsden judges that "McCosh's openness to at least a limited version of biological evolution helped the Princeton community weather the sensation over Darwinism. McCosh's views were too solidly rooted in the traditional Presbyterian paradigm and his personal presence too impressive for him to be easily labelled a heretic. At the same time he kept the Princeton faculty from being written out of the scientific community. By about 1875 this stance was particularly important strategically."[61]

McCosh would have been thrilled with that verdict. Reminiscing on twenty years as Princeton's president, he confessed that much of his time had been devoted to "defending Evolution, but, in so doing, [I] have given the proper account of it as the method of God's procedure, and find that when so understood it is in no way inconsistent with Scripture. I have been thanked by pupils who see Evolution everywhere in nature because I have so explained it that they can believe both in it and in Scripture."[62] He had already said as much in the preface to the Bedell Lectures, in which he reported that he had long "had a sensitive apprehension that the undiscriminating denunciation of evolution from so many pulpits, periodicals, and seminaries might drive some of our thoughtful young men to infidelity." If they saw evolutionary change "everywhere in nature" and yet were dogmatically told that "they could not believe in evolution and yet be Christians," the results would be disastrous. For himself, he was "gratified beyond measure to find that I am thanked by my pupils, some of whom have reached the highest position as naturalists, because in showing them evolution in the works of God, I showed them that this was not inconsistent with religion."[63]

Of course, Hodge occupied a different speech space. So, while "McCosh did not disagree with Hodge so much in principle as in emphasis,"[64] their modes of expression, as we have seen, were miles apart. McCosh's mission was to escort Princeton College into the brave new world of the research university; Hodge's passion was to preserve the Old School heritage of the Princeton seminary and keep it firmly tied to its traditional moorings. At the celebrations to mark the fiftieth anniversary of his own chair, Hodge reflected on his predecessors in precisely this vein. "Drs. Alexander and Miller were not speculative men. They were not given to new methods or new theories." Such conservatism delighted him. "I am not afraid to say," he went on, "that a new idea never originated in this Seminary."[65] Of course, this avowal cannot be taken at face value. For if Hodge was a solidly conservative thinker, he was no less a subtle one whose intelligence and influence were striking.[66] Nevertheless, his zeal to hold fast to tradition and to preserve his seminary as a site of doctrinal safekeeping placed him in a different space from that of his Scottish colleague.

Like Hodge, the distinguished Harvard botanist Asa Gray was fully aware of the drift away from Christian orthodoxy in the writings of many natural scientists. From what he called "the ultra-orthodox view of Dr. Hodge," a "long line of gradually-increasing divergence" could be traced in the work of men like William Thomson, John Herschel, Richard Owen, Alfred Russel Wallace, and Charles Darwin, on to the much greater radicalism of the likes of David Strauss, Carl Vogt, and Ludwig Büchner. In these circumstances, Gray insisted, it was critically important that the arrow of criticism was aimed with consummate care. In his review of *What is Darwinism?*, which appeared in the *Nation* on 28 May 1874 and argued for the coherence of a Christianized version of Darwinism, Gray concluded: "To strike the line with telling power and good effect, it is necessary to aim at the right place. Excellent as the present volume is in motive and tone . . . we fear that it will not contribute much to the reconcilement of science and religion."[67] To Gray, it seems, Hodge had taken up a position on the evolution question that, however laudable its love of theological fidelity, only succeeded in promoting exactly what McCosh wanted to avoid—the alienation of naturalists from their Christian heritage.

Evolution Princeton Style

In the years that followed, the way in which the theological fraternity at Princeton Seminary engaged with the Darwinian proposal was not unrelated to how their scientific counterparts in McCosh's college dealt with it. Something of the distinctive posture that the college adopted on the evolution question may be calibrated through the writings of several of its most notable science professors in the McCosh era. In 1854, Arnold Henry Guyot (1807–1884) took up the chair of geology and physical geography at the College of New Jersey. Swiss by birth, he had previously occupied a post as professor of history and physical geography at the College of Neuchâtel in Switzerland until its suspension by the Grand Revolutionary Council of Geneva during the revolution of 1848 which established the principality as a republic.[68] Following the urging of his friend and colleague Louis Agassiz, Guyot moved that summer to the United States and delivered a series of lectures at the Lowell Institute in Boston in 1849; these were later published under the title *The Earth and Man*.[69] This volume established his reputation in the English-speaking world and, after working for a few years for the Massachusetts board of education, he was appointed to Princeton, where he remained until his death in 1884.

Guyot's scientific reputation was built on his research on glacial motion, the laminated structure of glacial ice, and the character of erratic boulders, as well as on his work recording pioneering barometric measurements and finalizing plans for a national system of meteorological observation.[70] He had also studied theology at the University of Berlin and had embarked on the path towards ordination before turning more single-mindedly to geography and natural history. It is not surprising that, as an evangelical Presbyterian, he issued commentary from time to time on issues to do with science and religion and gave lectures on the subject to students at the Princeton Seminary and Union Theological Seminary in New

York. He also appeared at the 1873 meeting of the Evangelical Alliance in New York to speak at the session at which McCosh and Hodge were present.

Guyot's theme at that meeting was how to understand the creation narrative in the light of modern science. His purpose was to deliver a concordist reading of the first chapter of Genesis. He insisted that clinging to "an interpretation disproved by God's works" was folly and that it was imperative that the book of scripture should be read in the light of the book of nature. The literal twenty-four hour days of creation had to be reconceptualized. "To refuse, a priori, to believe in the possibility of this antique document agreeing in its teachings with modern science," he insisted, was itself profoundly "unscientific." So he sought to correlate the events of the biblical week of creation with scientific discoveries, all in a manner in keeping with the traditional concordism of the Christian geologists. Working up through what he called "the grand cosmogonic week described by Moses," he dealt with the creation of matter, the origins of the planets, and the appearance of plant and life forms during the Paleozoic, Mesozoic, and Quaternary epochs.[71]

What was particularly novel about Guyot's exegesis—and he had been developing these thoughts from as early as 1840, long before he moved to the United States—was the close correspondence he discerned between the Mosaic account and the nebular hypothesis.[72] To him, the formless waters that appear at the start of the narrative were none other than the gaseous matter of Laplace's theory of the origin of the solar system by means of slowly rotating, cooling nebulae. Guyot was sure that he had found here the key to unlocking the real scientific secrets of the Genesis creation story. He immediately set about "nebulizing" the narrative of the first three "days." On the first day, diffused gas was concentrated into dense clouds of dust, the following day witnessed the division of primitive nebulae into smaller nebulae, and finally came the concentration of these into stars and planets, including the Earth. Guyot had worked out all this more than thirty years earlier and had published his thoughts on the subject in the New York *Evening Post* in 1852, and spoken of them during the 1860s in lectures to Princeton and Union seminarians.

Intriguingly, despite Guyot's concerns to keep biblical hermeneutics in touch with scientific developments, neither Darwin's name nor the doctrine of evolution appeared throughout this 1873 account. This omission is perhaps not surprising, as Guyot had acquired a reputation as one of the vanishingly few remaining "creationist" scientists in North America. In 1880 when the editor of the *Independent* challenged the *Observer* to specify "three working naturalists of repute in the United States—or two (it can find one in Canada)—that is not an evolutionist," only two names surfaced. One was a Canadian, John William Dawson; the other was Arnold Guyot.[73]

Guyot's creationism was not static but suffused with developmental thought-forms, albeit of decidedly pre-Darwinian stripe. He was a profoundly organic—not mechanical—thinker, a perspective he had acquired through his encounter with continental romanticism in the form of Naturphilosophie and the direct influence of figures like Henrik Steffens, Johann Wolfgang von Goethe, and Karl Ritter. Accordingly he resorted frequently to organismic metaphors to explain natural and social phenomena rather than to the mechanistic images of Newtonian science. Nowhere is this sense of dynamic developmentalism more clearly seen than in his Lowell lectures, *The Earth and Man*. In these, he waxed lyrical in his identification of progress in the formation of the earth from Laplace's nebulae, the gradual specialization of organisms, the emergence of the continents, the progress of human civilization, and the like. He further explained in a letter to James D. Dana in 1856: "This law [of development] thus became for me the key for the appreciation and understanding and grouping of an immense number of phenomena both in Nature and History. My views of the human races and of universal history are, in great part, on the same base. So also the idea of the true sense of the first chapter of Genesis as a characteristic of the great organic epochs."[74] Thus while he never doubted the fixity of species boundaries or the need for divine action to originate life and propel development forward, his creationism was much less restrictive than Agassiz's inasmuch as he minimized divine creative intervention and allocated much to the action of natural law.[75]

For these reasons, it is not surprising that Dana could detect in Guyot's later 1884 *Creation* signs that he had been "led to accept, though with some reservation, the doctrine of evolution through natural causes." He noted that Guyot "excepted man, and also the first of animal life" from this process but insisted that "God's will was the working force of nature, and secondary causes simply expressions of it."[76] *Creation or the Biblical Cosmology in the Light of Modern Science* was only a moderately enlarged version of Guyot's Evangelical Alliance lecture, which finally appeared in print—at Dana's urging—the year of Guyot's death. In its closing pages, Guyot paused in his reflections on the "grand history of the creation" to "offer a few remarks on the relation it holds to evolution, the favorite doctrine of the day." And brief they certainly were. He simply insisted that evolution "from matter into life" and "from animal life into the spiritual life of man" was just "impossible," but he conceded that the "question of evolution within each of these great systems—of matter into various forms of matter, of life into the various forms of life, and of mankind into all its varieties—remains still open."[77] This was hardly an endorsement of McCosh's line, but neither was it an out-and-out rejection of transformism in favor of Agassiz-style fixism.

When Guyot let it be known that he wanted to retire from teaching geology, McCosh turned to the anti-Darwinian, Canadian geologist John William Dawson in hopes of luring him to Princeton. McCosh confided to him that the college's trustees had "set their hearts" on him. He offered Dawson a restricted teaching load and a salary equivalent to his own. The seminary joined in the appeal, offering Dawson a position there, too. Charles Hodge wrote warmly, listing the reasons why Dawson should respond positively to the call. The seminary's senior professor, William Henry Green, and Guyot himself also urged him to come. It was all in vain.[78] But the episode revealed that for McCosh, securing an advocate of what might be called doxological science was more important than finding someone who would endorse his own stance on evolution. When it came to the well-being of his beloved Princeton, he much preferred to appoint a pious anti-Darwinian than an unbelieving evolutionist. During the negotiations, McCosh lamented to Dawson that "if you do decline, which I hope you may not, we do not know where to look for a geologist of repute who is not a Darwinian."[79] Plainly for McCosh, evolution was not identical with Darwinism.

It was not the first time McCosh had tried to attract Dawson to Princeton. In 1871 he had put forward plans for the establishment of what would come to be known as the John C. Green School of Science at Princeton. They included a new chair of natural history, but it proved easier to formulate the proposal than to fill the post. The Smithsonian's Joseph Henry, a member of the appointment committee, suggested Edward Drinker Cope, a Quaker and an enthusiastic advocate of neo-Lamarckian evolution. McCosh did not feel comfortable with the proposal, perhaps because the board was wary about transmutationism, perhaps because he himself had reservations about the "biologization" of psychology.[80] Several other names were aired, including the zoologist Theodore Gill, the paleontologist Charles Frederick Hartt, and the geologist Joseph Le Conte, but those proposals came to nothing. The attempt to fill the chair dragged on and on. Later, Edwin G. Conklin attributed the difficulties to the anti-Darwinian sentiments that Hodge's *What is Darwinism* had stirred up.[81]

In the end, McCosh's eye turned back toward Ireland to a former student, George Macloskie (1834–1920), like himself an ordained Presbyterian minister with a degree in philosophy and interests in natural history. Back in 1871, Macloskie had written to McCosh, expressing his interest in finding a career, ministerial or academic, in the New World. His old teacher replied, "I had heard of your honors in a general way but not specifically and I am pleased to find that you remember me and give me the details of yourself." "There would be fine fields of usefulness for you in America," he went on, "but the difficulty would be to get you a stranger introduced to them . . . The difficulty is to get you known. I give no advice. The responsibility must be with yourself. But I will be glad to hear from you as to your views and intuitions."[82] Now, after several scuttled efforts to fill the natural history chair, Macloskie came to McCosh's mind. After all, he walked the narrow path between the materialism and the idealism that other candidates seemed incapable of negotiating. Although he can hardly have been one of the newly professionalized breed of scientists that McCosh craved for his college, Macloskie's stance on evolution mirrored McCosh's to perfection; in the circumstances, that counted for much.[83]

Macloskie affected to be surprised to receive the invitation from his old Belfast professor, with whom he had also fought in support of Gladstone's Disestablishment policies.[84] It is reported that he said he only accepted the position because "he wanted to see America for a few years and thought he might learn enough natural history on the way over to keep ahead of the boys."[85] This remark is not to be taken at face value. He certainly had the paper credentials for an academic position. At Queen's College Belfast he received a gold medal for undergraduate work in natural history; there he had taken classes with such scientific luminaries as McCosh's coauthor George Dickie, an expert on marine algae and a Fellow of the Royal Society; the physical chemist Thomas Andrews, collaborator of P. G. Tait and an outstanding classicist besides; and the Scottish zoologist Sir Wyville Thomson, who later served as chief naturalist on the famous *Challenger* expedition. Macloskie, evidently a polymath, also undertook theological training and while serving as a Presbyterian minister between 1861 and 1873 pursued legal studies— for which he also received a gold medal—as an external student at the University of London. He even sported a letter, dated 14th March 1865, from the vice president of Queen's, Thomas Andrews, lauding the young man's linguistic competences to qualify him for a chair of Greek and Latin, and he carried a similar 1868 recommendation about his scientific accomplishments from George Dickie, his former Queen's professor who had gone on to hold the chair of botany at Aberdeen.[86] Prior to crossing the Atlantic, Macloskie had already made some tangential observations on the Darwinian theory in an article he wrote in 1862, the year after his ordination, entitled "The Natural History of Man." At that point he was hesitant about the universal claims of some Darwinians, but it did not deter him from affirming that natural selection could account for human racial differentiation. Indeed, he claimed that he himself had "already employed this principle, to explain the diversity that exists in the different tribes of mankind, whilst the specific unity is still preserved."[87] Admittedly the deductive character of the Darwinian theory was, at this stage, bothersome to Macloskie, though in later years he came to defend the necessity of scientific speculation. But his early stance already displayed some of the qualities that McCosh himself would espouse, namely, openness to evolution's empirical findings and resistance to its materializing inclinations.

From the time Macloskie arrived to take up the chair of biology in 1875 until his retirement in 1906 and indeed beyond, he vigorously promoted the virtues of science among orthodox Presbyterians. For more than thirty years he practiced doxological science, rounding off his lecture notes on the protozoa by affirming that the "study of Natural History forces on the mind the conviction that there is knowledge and wisdom and a controlling mind behind these things. No mind is so strong in its naturalism as to get over this conviction and recent discoveries are proving more than ever that whatever share secondary causes and accidents have had in making things what they are, behind and above all there is some intelligent one who is directing them all according to the counsels of His own will."[88] Macloskie was exactly what McCosh needed—a colleague who could mount an assault on materialism while retaining a belief in the fact of evolutionary change[89]—and he was prepared to sacrifice the professionalizing aspirations he had for the college, even if only as an interim measure, to preserve theological integrity.

Macloskie himself recognized that even though he produced *Elementary Botany* in 1883, prepared the botanical sections of the three-volume *Reports of the Princeton Expedition to Patagonia,* and wrote on the housefly for the *American Naturalist,* his "discussion of the hot questions of science and the Bible" was his "most useful work."[90] Over the years he took on something of the mantle of keeping the Presbyterian community in touch with scientific developments through his numerous publications in church-related serials. During the 1880s and 1890s, he called their attention to the role of speculation in science, accommodationist strategies, and theistic evolution. In 1887, for example, he defended not just the legitimacy but the necessity of speculative theorizing in science—a not insignificant twist of argument, given the nervousness about theory emanating from his own community's deep attachment to the Scottish philosophy. Subverting standard Baconian induction, he went so far as to suggest that "hypotheses that conflict with recognized principles of science, of philosophy, or of theology are all legitimate subjects of examination."[91] Plainly, scriptural inspiration did not mean that the laws of nature could be deduced from incidental biblical references to physical phenomena. "We are not at liberty" he remarked, "to erect our science

upon the Scripture, and then to turn round and prove the Scripture by our science."[92] Thus, he declared scriptural geology a dead force and railed against the
strained efforts of the concordists to "geologize" Genesis. What such wooden
harmonizing schemes ignored was the old Calvinist doctrine of scripture's "accommodating itself" to conventional human talk.[93] Indeed, that principle provided a strategy for coming to terms with Darwin's theory with poise and patience.
To follow the creationist science of Louis Agassiz—"not a theologian and scarcely
a Christian"[94]—would be nothing but folly. Besides, Macloskie insisted that scientists should not be held responsible for the inferences that others made, that
they had "the right to go wrong" and when they did so should not be charged with
"moral delinquency," and that "erroneous assumptions" could turn out to be "a
necessary step" in scientific progress. What really bothered him was the readiness of too many theological types—and he likely had Woodrow's enemies in
mind[95]—to resort to talk of science's "evil tendencies." "Men brand unwelcome
doctrines as having an evil tendency," he quipped, "when they see no direct answer to them."[96]

Crucial to Macloskie's own domestication of Darwinian declarations was his
belief that he had satisfactorily resolved the issue of teleology. He repeatedly
spelled out his resolution to his readers. Evolution was itself a goal-oriented operation. For at least a decade, he had been greedily fastening on Huxley's admis-
sion that while Darwin's theory "does abolish the coarser forms of Teleology, it
reconciles Teleology and Morphology. The teleological and the mechanical views
of nature are not mutually exclusive."[97] This idea opened the door to a chastened
form of theistic evolution, and in an 1898 essay replete with testimonial gobbets
from figures like John Fiske, Frederick Temple, Asa Gray, and James McCosh,
Macloskie reiterated his belief in evolutionary teleology. Not that everything was
reducible to natural law. "The Christian evolutionist," he urged, "is able to believe
in the miraculous birth, the theanthropic personality of Jesus Christ, in His
miracles, in His resurrection after death and in His ascension to glory."[98] None
of these beliefs, he was convinced, was incompatible with evolution in general or
human evolution in particular. Like St. George Mivart, Macloskie thought it entirely conceivable that the human physical form was the result of a naturalistic
evolutionary process, even if the "spiritual nature" was the consequence of special divine intervention. "Evolution, if proven as to man," he announced, "will
be held by the biblicist to be a part, the naturalistic part, of the total work of his
making, the other part being his endowment miraculously with a spiritual nature . . . As a member of the animal kingdom man was created by God, probably
in the same naturalistic fashion as the beasts that perish; but unlike them, he has

endowments which point to a higher, namely a supernaturalistic, order of creation."[99] He was sure that the monogenetic implications of the Darwinian theory had actually done much to confirm the biblical narrative. With these convictions, it is not surprising that Macloskie sided with Gray over against Hodge when he commented, "We have no sympathy with those who maintain that scientific theories of evolution are necessarily atheistic."[100] Moreover, he insisted that attempts to make evolution a heresy were disastrous for religion and science alike, and he applauded McCosh and Gray for their "great service" in demonstrating that evolution "is not dangerous."[101] All of these convictions were entirely in keeping with McCosh's vision. So, from the vantage point of the early twentieth century, he looked back with appreciation to McCosh's early efforts to Christianize evolution and rejoiced that McCosh's scheme, once the subject of sharp criticism, was now being "universally accepted" as an explanation "for the creation of new species."[102]

Guyot's half-hearted, anti-Darwinian developmentalism and Macloskie's more strenuous endorsement of evolution were alike premised on the presumption of an internal telos intrinsic to the globe's material and biological history. It was the same for Hodge and McCosh. Hodge's insistence that Darwinism was inherently subversive of design and McCosh's confidence that evolutionary history was nothing less than the outworking of in-worked purpose resulted in markedly different rhetorical stances on the whole subject. For Darwin's Princeton friends and foes alike, teleology was what mattered most. And it is notable that in the following generation, some of Presbyterian Princeton's most celebrated students of the earth sciences continued to promote versions of evolution that bore a strongly teleological imprint.

The eminent vertebrate paleontologist William Berryman Scott (1858–1947)—second John I. Blair professor of geology from 1886 to 1930, student of and successor to Guyot, and grandson of Charles Hodge—is a case in point. As a sixteen year-old, he had corrected the proofs of *What is Darwinism?* for his grandfather. At that stage he shared Hodge's conclusions, describing himself as an "ardent anti-evolutionist," but he later came to side with McCosh, who conferred with the younger man when composing *The Religious Aspect of Evolution* in 1887. Nevertheless, Scott was keen to point out in his autobiography that Hodge retained great affection for Darwin's writings till the end of his days. As he recalled it: "In the spring of 1878, my Grandfather's health began to fail, but did so very gradually. One day he said to me: 'I wish you would get me Darwin's *Voyage of a Naturalist* from the library.' I brought it to him and he read it with the keenest interest. When he handed the book to me to return it, he said: 'That is a very remarkable and delightful book; now get me Wallace's *Malay Archipelago*.' When he had read that, he said: 'That's an excellent book, but not to be compared with Darwin's.' "[103]

In the interim, Scott had come into the orbit of Thomas Henry Huxley, with whom he studied in London during the time he and Henry Fairfield Osborn traveled together as students in Europe between 1878 and 1880. Besides the reputation he acquired early on for building up the university's fossil collection, particularly of Ogliocene mammals, Scott played a major role in the development of the earth sciences at Princeton and served from 1918 to 1925 as president of the American Philosophical Society, the country's most august learned body. Throughout his entire career he conducted research on the evolution of life on earth. But as his obituarist G. C. Simpson observed, "he stressed the evidence for the reality of organic evolution more than the theoretical principles as to how and why evolution has occurred."[104] Early on, he set his face strongly against the theories of August Weismann, whom he dubbed the "Pope of the Neo-Darwinists," and adopted a definite, though not extreme, neo-Lamarckian stance—a standpoint entirely in keeping with what he had gleaned from McCosh. He continued to be attracted to the idea of goal-directed evolutionary development, which fostered his interest during the 1890s in Wilhem Waagen's ideas about predetermined variation. And it bothered him that, as he quipped, the "creed of the Neo-Darwinists was: 'There is no god but natural selection and Darwin is its prophet.' "[105] He was thus convinced, according to Simpson, that "random and discontinuous variations . . . have little to do with evolution, which normally proceeds by continuous and oriented change." Scott plainly remained a convinced evolutionist of non-Darwinian stripe and during the decades when numerous—and, in Simpson's eyes, sometimes bizarre—alternatives were gaining converts,[106] Scott "saw the difficulties and it is wholly to his credit that his intellectual honesty and his faith in objective science prevented his entering the false paths followed by so many of his colleagues."[107]

Nowhere, perhaps were these perspectives more clearly articulated than in Scott's 1914 Westbrook Lectures at the Wagner Free Institute of Science in Philadelphia. These later appeared, in 1917, under the title *The Theory of Evolution*. Acutely aware that he was writing at a time when, well in advance of the later neo-Darwinian synthesis, evolutionary theory was under attack from numerous scientific quarters, he reaffirmed its fundamental character with an inaugural announcement:

> It is widely believed that the theory is an outworn device, which naturalists are beginning to discard and that soon it will have a merely historical interest. This mis-

understanding, for such it is, has arisen from the debates among zoologists and botanists as to the manner in which evolution has actually occurred and the efficient causes which have brought it about, and, further, from the ambiguous way in which the term "Darwinism" is often employed. Frequently, the term is made a synonym of evolution, but it ought properly to be restricted to Darwin's *explanation* of evolution by natural selection.[108]

The situation in early twentieth-century Darwinian theory thus allowed Scott the flexibility to retain his skepticism about the all-sufficiency of natural selection while proclaiming a robust confidence in the evidence for evolution itself. "Personally, I have never been satisfied that Darwin's explanation is the rightful one," he wrote. "The doctrine, in its application to concrete cases, is vague, elastic, unconvincing and seems to leave the whole process to chance." To Scott, there was nothing at all odd that this declaration should appear in the early pages of a volume detailing evidence for the reality of evolution. Disagreement about the mechanism of transformism was one thing; rejecting descent with modification entirely another. After all, as Scott did not hesitate to point out, the doubts that figures like Edmund Beecher Wilson and William Bateson entertained about the causes of evolution did not mean they were skeptical about its reality. "Naturalists are all but unanimous in accepting the theory of evolution as an established truth," he announced; at the same time, "there is every possible divergence in their views as to the causes of development and diversification."[109] It was precisely this circumstance that enabled Scott to adopt different theories at different times—neo-Lamarckism, orthogenesis, and the like—without abandoning the fundamental principle of evolutionary change.

The sense of evolutionary directionalism that pervaded McCosh's thoughts on the subject and wove its way through the contributions of his student William Berryman Scott also surfaced in the writings of another distinguished son of Princeton, Scott's friend, colleague, and coauthor Henry Fairfield Osborn (1857–1935), who later became director of the American Museum of Natural History in New York.[110] He, too, spent his formative years steeped in the social and religious environment of Scots Presbyterianism and studied geology with Guyot at Princeton, where he absorbed from McCosh the idea of evolution as God's method of creation. According to Osborn's student, colleague, and obituarist William King Gregory, McCosh played "a great part in making Osborn also primarily a philosopher and teacher who consistently fought for 'idealism' in opposition to 'materialism' of the type exemplified by Ernst Haeckel."[111] A sense of predetermination by heredity and a skepticism about chance which characterized Osborn's own autobiographical reflections no less pervaded his sense of evolutionary history. Like Scott, he found himself attracted to the neo-Lamarckian strand of evolution championed by Edward Drinker Cope, with whom he later collaborated and whose biography he wrote, even if Osborn later sided with Weismann's theory of the germ plasm.

Osborn's enthusiasm for evolution manifested itself in many ways, not least at the American Museum of Natural History. But nowhere was it more obviously on display than in his involvement in the build-up to the infamous 1925 Scopes Monkey Trial in Dayton, Tennessee.[112] A year or two earlier, Osborn had written to the *New York Times* to contest William Jennings Bryan's claims about the lack of evidence for human evolution. That catapulted him into the center of the debate, and he emerged as the chief scientific adversary of Bryan, the Great Commoner who was three-times Democratic candidate for the U.S. presidency. Later, in *Evolution and Religion,* a tract for the times that Osborn issued soon after the trial, he announced that in Bryan's case "religious fanaticism" had simply drowned out scientific reason.[113] For all that, Osborn made it clear in the *New York Times* that "evolution by no means takes God out of the Universe, as Mr Bryan supposes."[114] As Osborn's brother later pointed out, "The whole tenor of his nature was towards creation and one of his last views on evolution was, if I correctly understood it, a belief that life contains within itself a creative power which leads it to adapt itself to its external surroundings and to create such new forms for itself as are required for its existence. This seems to me to be a distinctly religious conception as opposed to the mechanistic theory of straight survivalism by the selection of the fittest under the long processes of trial and error."[115] Gregory concurred, adding that Osborn, "as the disciple of Dr. McCosh, consistently regarded evolution as an expression of the 'firm and undeviating order' conceived by the divine creative mind."[116]

Macloskie, Scott, and Osborn all shared a strongly evolutionary conception of nature that bore the stamp of McCosh's vision and, to some degree at least, Guyot's pre-Darwinian developmentalism. Their collective emphasis on the directional nature of evolutionary change, coupled with a firm conviction in the unassailability of the evidence for evolution, constituted a rather distinctive Princetonian take on the whole subject. What they promoted was an openness to non-Darwinian, linear forms of evolutionism that allowed for the operation of a range of transformative mechanisms above and beyond random variation and pan-adaptationism. Ronald Rainger has commented on this distinctively Princeton ethos, noting that Scott's "conception of evolution in definite directions met with McCosh's wholehearted approval," and adding that Osborn, drawing "on the collaborative, religious framework for scientific inquiry at Princeton . . . developed an evolutionary theory that reflected McCosh's influence and amplified Scott's ideas."[117] In promoting this perspective, these figures cultivated links with Edward Drinker Cope, Alpheus Hyatt, and other like-minded scientists who expressed increasing doubts about the all-sufficiency of natural selection in the decades before the neo-Darwinian synthesis began to take form in the late 1930s. How the theologians across the campus in the seminary reacted to the Darwin phenomenon during the later decades of the nineteenth century could not, and did not, remain untouched by how these local scientific champions conducted their conversation with the new biology.

Calvinizing Evolution

The orthogenetic versions of evolution strongly promoted at Princeton University during the decades around 1900 resonated more with the sentiments of the theologians at the seminary than other less directional evolutionary schemes. Nowhere, perhaps, was this more conspicuously manifest than in the case of B. B. Warfield.[118] We have already noted how he described himself as a "Darwinian of the purest water" at the time of McCosh's arrival at the college. Later, he reflected, he "fell away from this, his orthodoxy." When McCosh informed him "with some vigor" that "all biologists under thirty years of age were Darwinians," Warfield mused that he himself was "about that old" before he "outgrew it."[119]

Warfield penned these words in 1916, just a year after he had brought out a lengthy article, "Calvin's Doctrine of the Creation," in the *Princeton Theological Review*. Here he made much of Calvin's insistence that the term "creation" should be strictly reserved for the initial creative act. For the subsequent "creations" were—technically speaking—*not* creations ex nihilo but rather modifications by "means of the interaction of its intrinsic forces." The human soul was the only exception, for Calvin held to the creationist (as opposed to the traducianist) theory that every human soul throughout the history of propagation was an *immediate* creation out of nothing. To Warfield, then, Calvin's doctrine opened the door to a controlled "naturalistic" explanation of natural history—including the human physical form—in terms of the operation of secondary causes. These, to be sure, were directed by the guiding hand of providence, but this conviction did not prevent him from asserting that Calvin's doctrine of creation actually turned out to be "pure evolutionism." In addition, while acknowledging that Calvin himself had understood the days of creation as six literal days, he nonetheless believed that Moses, "writing to meet the needs of men at large, accommodated himself to their grade of intellectual preparation, and confine[d] himself to what meets their eyes." Now, in the twentieth century, Warfield suggested that in order to perpetuate the spirit of Calvin's hermeneutic genius, "it was requisite that these six days should be lengthened out into six periods—six ages of the growth of the world. Had that been done Calvin would have been a precursor of the modern evolutionary theorists. As it is, he only forms a point of departure for them to this extent—that he teaches, as they teach, the modification of the original world-stuff into the varied forms which constitute the ordered world, by the instrumentality of second causes—or as a modern would put it, of its intrinsic forces." All this meant that "Calvin very naturally thought along the lines of a theistic evolutionism."[120] Whatever Warfield intended by saying that he abandoned McCosh's Darwinian orthodoxy by the age of thirty, this declaration certainly cannot be taken to mean that Warfield believed that the idea of evolutionary transformism was anything but fully compatible with Calvinist orthodoxy.[121]

Bearing in mind the reservations about the omnipotence of natural selection as an evolutionary mechanism that were circulating among prominent Princeton scientists at the time, Warfield's observations are entirely understandable. An avid reviewer of scientific works, he was acutely aware of what Peter Bowler calls the "eclipse of Darwinism" in the fin de siècle scientific world when a variety of non-Darwinian evolutionary theses flourished—notably, neo-Lamarckism, orthogenesis, and mutationism.[122]

But there was another source of hesitancy about Darwinian absolutism that for Warfield lay literally closer to home—animal breeding. The Warfield family's investment in the breeding of shorthorn cattle and William Warfield's signal contributions to the American literature on the subject have already attracted our attention. So, too, has the approving eye that William cast on Darwin's contributions to this subject. But animal breeding also involved practices troubling the idea that the selection of accidental variations was the only factor in the production of new, stable breeds. As Bert Theunissen has shown, animal breeders knew all too well about the critical role of careful inbreeding and crossbreeding in the production and stabilization of new varieties. "In-and-in breeding," as it was called, was needed to "fix" and preserve the features of a new breed, and it required the most careful oversight involving occasional out-crosses to prevent degeneration and retain vigor. Stock breeders, of course, were notoriously secretive about their practices, and in the world of animal husbandry it was exceptionally difficult, as Darwin himself discovered, to figure out just who was trustworthy. Nevertheless, if Theunissen is correct, Darwin downplayed inbreeding and crossbreeding in order to promote his central analogy between artificial and natural selection. The anti-Darwinian Scots Presbyterian John Duns blew the whistle in a negative review of the *Origin* in 1860: "Has man's intelligence gone out in seeking variation by selection only?"[123] No, Duns insisted. "Cross-breeding, and breeding in-and-in, under man's watchful care and discriminating intelligence, can alone give the key to variation."[124] In Darwin's case, Theunissen suggests that pigeon breeders, from whom Darwin gleaned so much of his information, were atypical in the breeding world and that he was "eager to take the pigeon fanciers at their word, as it was only their methods that provided him with the perfect analogy with natural selection."[125] But the world of shorthorn cattle breeding was different, and the Warfields knew all about the strengths and weaknesses of inbreeding, linebreeding, out-crossing, and so on.

The very chapter in which William Warfield laid out the principle of what he called "Nature's selection," significantly, was entitled "Inbreeding." Pointing out that, in general, natural selection tended toward the maintenance of the "average," he immediately moved on to a discussion of how human intervention through breeding management had worked against nature's conservatism. Frequently the stabilizing of "superior qualities" had been secured through "highly incestuous breeding." Reviewing the work of "Robert Bakewell, the celebrated improver of Leicestershire sheep and Longhorn cattle," William described how he had used the system of inbreeding to great effect. But for long enough, he noted, a "veil of mystery was thrown over most of his proceedings" for in fact Bakewell had found it necessary to engage in "skillful interpositions of remote affinities when he saw or apprehended danger." As for the breeding of "toy" pigeons, William was sure that the "tiny size" could only be "maintained by the most constant return to a single line of blood—mating brother and sister, and similar cases of incestuous crosses." For all that, William found the practice objectionable. Among humans, he recalled, "divine law" forbade "incestuous marriages" and with good cause; "idiocy, insanity, consumption, and scrofula" had all "resulted from a defiance of this law." Ultimately, inbreeding brought deterioration, and that was why those practicing it had to maintain constant vigilance. His own preference was for natural breeding; it was the method that delivered the best long-term results "without any risk of doing injury to the breed of cattle which is cultivated." Followers of "evanescent fashions"—fancy pigeon breeders doubtless among them—might pursue other systems, but for "the seeker for safe and plain methods, the system of natural breeding will continue in the future to hold the first place."[126]

Given his family background and first-hand experience of the stockyard, Benjamin Warfield was in a position, unique among the Princeton theologians and doubtless much further afield, to ascertain just how controverted questions about inheritance, breeding, variation, reversion, and artificial and natural selection really were. After all, the fifty-three pages of clippings he kept of his father's contributions to the culture of shorthorns included the elder Warfield's reviews of the latest books on the science of breeding—books like J. H. Sanders' *Horse-Breeding; Being the General Principles of Heredity*, which he had in his personal library and which examined in detail Darwin's and Spencer's thinking on inheritance. Here, William recorded his own views on subjects like atavism, the environmental causes of variability, the fixing of accidental variations, and, crucially, his conviction that "inbreeding is always an evil . . . but in the formation of a breed it is a necessary evil and need not produce bad results if very carefully watched."[127] William thus provided a kind of breeder's theodicy for the practice; inbreeding was unavoidable if any desirable trait was to lead to a new strain.

His father's example taught Benjamin Warfield the importance of keeping abreast of scientific opinion and retaining an open mind on the latest theory in natural history. The sense of prevarication—even disquiet—that Warfield exhibited in the decades between his early profession of Darwinian faith and his later willingness to evolutionize Calvin's theology of creation, reflect his awareness of changing scientific theories on non-Darwinian evolutionary mechanisms every bit as much, if not more, than his theological predispositions. His reading of Vernon Kellogg's celebrated *Darwinism Today*, for example, fully exposed him to—as the volume's subtitle expressed it—the latest "discussion of present-day scientific criticism of the Darwinian selection theories, together with a brief account of the principal other proposed auxiliary and alternative theories of species-forming." Kellogg (1867–1937) was an entomologist and evolutionary biologist who was deeply critical of pan-Darwinian selectionism. Though Warfield took a dim view of Kellogg's understanding of teleology, he warmly welcomed the volume because of its careful denunciation of all-sufficient ultra Darwinism and because it gave voice to the plethora of competing speculations circulating in evolutionary circles.[128]

Warfield's reservations about Darwinism as a comprehensive explanatory system had begun to manifest themselves by the time he moved from the seminary in Pittsburg to Princeton in 1887. The following year he began delivering his "Lectures on Anthropology," which included the subject "Evolution or Development." Here he made it clear that Darwinism did not enjoy anything like the scientific status of the Newtonian system, for there were just too many unresolved geological, zoological, and paleontological anomalies—anomalies he did not hesitate to itemize. Recapitulationist arguments drawn from embryology on the parallels between the development of the embryo and the evolution of species, of the kind beloved by Ernst Haeckel, particularly irked him, and indeed they went into rapid decline around 1900.[129] Evolution, he insisted, quite simply was not yet proven—an opinion he repeated in a review of McCosh's *The Religious Aspect of Evolution*.[130] Besides, he was sure that Christianity was incompatible with "a thoroughgoing evolutionism." And yet, aside from the narrative of the creation of Eve, which he described as "a very serious bar in the way of a doctrine of creation by evolution," Warfield could announce, "I am free to say, for myself, that I do not think that there is any general statement in the Bible or any part of the account of creation, either as given in Genesis 1 and 2 or elsewhere alluded to, that need be opposed to evolution." Nor did his uncertainties about evolution's empirical robustness prevent him from concluding the lecture: "The whole upshot of the matter is that there is no *necessary* antagonism of Christianity to evolution, *provided that* we do not hold to too extreme a form of evolution." If the constant supervision of divine providence and the "occasional supernatural interference" of God were retained, he concluded, "we may hold to the modified theory of evolution & and be Christians in the ordinary orthodox sense. I say we may do this. Whether we ought to accept it, even in this modified sense is another matter, & I leave it purposely an open question."[131] If Warfield's sense of equivocation is evident here, so too is the openness of his stance and his insistence that the theory should be tested ultimately by its empirical adequacy.

That same year Warfield published a lengthy and sympathetic account of Darwin's life and letters. In particular, he focused on Darwin's loss of faith and was at pains to show that his spiritual pilgrimage was a needless misfortune. In sketching what he called Darwin's "spiritual biography," Warfield paused to comment: "We raise no question as to the compatibility of the Darwinian form of the hypothesis of evolution with Christianity. Mr Darwin himself says that 'science' (and in speaking of 'science' he has 'evolution' in mind) 'has nothing to do with Christ, except insofar as the habit of scientific research makes a man cautious in admitting evidence.'" At the same time, Warfield was convinced that Darwin's construal of teleology and the presumption that his theory destroyed any confidence in divine design were untenable. To him, Darwin's conception of design was particularly confused, and he reviewed in considerable detail the correspondence on the subject between Darwin and Asa Gray. At base Warfield, like Gray, thought Darwin's understanding of design much too crude in positing "a single extrinsic end as the sole purpose of the creator," and he insisted that "no one would hold to a teleology of the raw sort which he here has in mind." Not at all surprisingly, Warfield welcomed the way Alfred Russel Wallace "showed that there was no more difficulty in tracing the divine hand in natural production through the agency of natural selection, than there is in tracing the hand of man in the formation of the races of domesticated animals through artificial selection."[132]

Warfield continued to keep an interested eye on literature, both scientific and theological, bearing on evolutionary questions in the years that followed. He reviewed a wide range of such works in an effort to keep the readers of his own, newly established *Presbyterian and Reformed Review* up to date. The writings of McCosh, William Henry Dallinger, Henry Calderwood, James Iverach, John George Romanes, Vernon Kellogg, Ambrosius A. W. Hubrecht, and a host of others all came under his inspection. What clearly emerges from these and other writings is that while Warfield protested the theological propriety of a properly conceived evolutionism—even if it sometimes fell far short of full empirical verification—he did not share the thoroughgoing enthusiasm for the theory that his Scottish counterparts voiced. While he found much to commend in Calderwood's *Christianity and Evolution,* he welcomed several critical readings of Henry Drummond's contributions and recorded that he was "by no means as sure as is Dr. Iverach of the reality of evolution in the wide range which he gives it."[133] And so by 1895 in a survey, "Present Day Conception of Evolution," he insisted that it was simply "a more or less probable or more or less improbable" working hypothesis and remained "as yet unproved."[134]

Even so, in his newly minted journal, Warfield went out of his way to expose readers to serious engagement with the latest evolutionary thinking. In its inaugural year, 1890, he organized a symposium on the whole question of the nature of animal life, drawing together a range of prominent scholars.[135] Among them was William Berryman Scott, whose observations were particularly striking as he contended for the possibility of an evolutionary account of the human soul with a suite of arguments that Bradley Gundlach suspects Warfield and Scott had already discussed in private conversation.[136] That Warfield remained committed to the creation of the soul ex nihilo did not prevent him from speculating that it might "not matter where we place the point of interference" that introduced into nature the "immaterial substance" that is the "soul"—whether "at the apparition of man, or at the influx of animal life, or at the beginning of life itself."[137]

The evolution question continued to absorb Warfield as the new century dawned. In 1901 he issued a subtle analysis of "Creation, Evolution, and Mediate Creation." Here he carefully discriminated between creation out of nothing and evolution as the modification of material through intrinsic forces. Plainly, these were two entirely different operations. Mediate creation he considered as moments of direct divine intervention "in the course of his providential government by virtue of which something absolutely new is inserted into the complex of nature"—a move designed to retain the integrity of the miraculous. Warfield, of course, was painfully conscious that there were those pushing evolutionism far into anti-supernaturalist territory, but for all that, he insisted, "the Christian man has as such no quarrel with evolution when confined to its own sphere as a suggested account of the method of the divine providence."[138]

Two years later, in an examination of human origins, he was sure that "the biblicist is scarcely justified in insisting upon an exclusive supernaturalism in the production of man such as will deny the possibility of the incorporation of natural factors into the process." Warfield certainly allowed the possibility of "an evolutionary process in the production of man," provided room was left for some direct creative act of God in the operation.[139] This was *precisely* what he understood Calvin to teach. For when he turned to Calvin's theology of creation in 1915, he cast Calvin's account of human origins in these very terms. The human soul, he observed, was—and continued to be for every human being—a direct act of creation ex nihilo, but the human body was the product of intrinsic or evolutionary forces. As Warfield summarized it: "he ascribed to second causes . . . the entire series of modifications by which the primal indigested mass called heaven and earth has passed into the form of the ordered world which we see, including the origination of all forms of life, vegetable and animal alike, inclusive doubtless of the bodily form of man. And this, we say, is a very pure evolutionary scheme."[140] On at least half a dozen occasions, Warfield described Calvin as an evolutionist or his theology of creation as evolutionary. Whatever his own sense of evolution's empirical ambiguities and whatever his awareness of the competing and often conflicting alternatives to totalitarian Darwinism, Warfield's rhetoric was entirely remarkable for its efforts to evolutionize Calvin and to Calvinize evolution. For a high Calvinist like Warfield, the theological legitimacy of evolution could scarcely have had a more august ancestor.

Of all the Princeton theologians in the generation after Charles Hodge, Warfield, the "Lion of Princeton,"[141] made the most sustained contribution to assessing evolutionary theory. Others also expressed views on the subject from time to time. In general, their evaluations were in keeping with Warfield's diagnosis. Their policy was to insist that Calvinist theology could readily absorb evolution properly circumscribed. Less wholehearted than their Scottish counterparts but decidedly more forbearing than their fellow Presbyterians in the South, they developed a kind of anticipatory apologetic laying out the path to be taken should the theory achieve increasing empirical corroboration. A couple of brief instances will suffice to give a flavor of their plan of attack.

Preemptive diplomacy was exactly what Joseph van Dyke had in mind when he published *Theism and Evolution* in 1886. It was also what Archibald Alexander Hodge (1823–1886), who introduced the volume, plainly endorsed. He was sure that "when strictly confined to the legitimate limits of pure science this doctrine of evolution is not antagonistic to our faith as either theists or Christians." When construed anti-teleologically, he insisted that Charles Hodge, his father and predecessor in his chair of theology, had been "abundantly justified in indicating this phase of evolution as atheistic." But, he went on, "Evolution considered as the plan of an infinitely wise Person . . . can never be irreligious." Of course, he did not mean that the theory could yet be accepted as empirically validated. As he explained, "It is not intended in all that has been said to express any opinion as to the truth of evolution in any of its forms, but only to indicate the limits, on the respective sides of which Christians, as such, can have no controversy, or no truce." The real enemy was "not evolution as a working hypothesis of science dealing with facts, but evolution as a philosophical speculation professing to account for the origin, causes, and ends of all things."[142] In fact, he had been saying similar things for quite some time. By the time a rewritten version of his *Outlines of Theology* appeared in 1878, the leading scientific opponent of Darwin in America, Louis Agassiz, was now gone, and support for the theory was being mounted by theologically astute scientists like Asa Gray, James Dana, and George Frederick Wright. The opposition to evolution prominently featured in the first printing had diminished, to be replaced by the concession that the natural theologian had, "of course, only the most friendly interest" in evolutionary theories that did not deny teleology.[143] Shortly afterward, in a review of Asa Gray's *Natural Science and Religion*, the rhetoric A. A. Hodge adopted markedly diverged from his father's when he described Gray as "a thorough evolutionist of the Darwinian variety, and at the same time a thoroughly loyal theist and Christian." Right from the outset, he made it clear that he "had no sympathy with those who maintain that scientific theories of evolution are necessarily atheistic." Again his judgment was that a progressivist evolutionism that allowed for "a providential unfolding of a general plan" was far from theologically objectionable.[144] He was even prepared to entertain the possibility that the human body had undergone some evolutionary transformation, and he found a good deal to commend in Henry Drummond's *Natural Law in the Spiritual World*—a work that had "undoubtedly assisted in opening up a vein of important truth."[145]

These judgments not only echoed the direction in which scientific thinking at Princeton College was moving; they resonated with the ways in which Warfield was engaging with post-Darwinian evolutionary theory. In fact, Warfield, who would later succeed to A. A. Hodge's chair at Princeton, joined forces with him to produce an influential article on the nature of biblical inspiration in 1881 which served, in the shadow of the Robertson Smith affair in Scotland, as a crucial element in the conservative critique of the liberal views of Charles A. Briggs, Henry Preserved Smith, and Samuel I. Curtiss. During the following decade, as conservatives worried more and more about the radicalism of the new biblical criticism, heresy charges were brought against Briggs, who was eventually suspended from the ministry.[146] But Benjamin Warfield and Alexander Hodge also discussed the evolution question. When Hodge found himself mobilized in support of James Woodrow in the controversy among the Southern Presbyterians, he confessed to Warfield in 1884, then still teaching in Allegheny Seminary, that he had no desire to "be mixed up in any way" in that scuffle.[147] He made a clean breast of what he considered evolution's empirical fragility and worried over the creation of Eve in any human evolutionary scenario. In public, he pronounced on evolution's compatibility with theology; in private, he doubted that the evidence was yet conclusive. Already embroiled in controversy over the nature of biblical inspiration, his strategy was anticipative and, in some measure at least, intended to prevent any blow-up comparable to the storm battering the Southern Presbyterians.

The Bermuda-born Francis Landey Patton (1843–1932) adopted a similar apologetic strategy. He served as "Professor of the Relation of Philosophy and Science to the Christian Religion" at Princeton Seminary from 1881 until his appointment as President of the College of New Jersey in 1888.[148] In 1902, he returned to the seminary as president. Holding steadfastly to Old School values, he sided with the fundamentalists in their disputes with modernism during the early years of the twentieth century. Earlier, just before his arrival in Princeton, he had reflected on the William Robertson Smith affair in Scotland. In a relatively measured diagnosis, he made it clear that what was concerning about the whole affair was "the prevailing tendency" that the spread of higher critical philosophy induced. To Patton, that inclination was infinitely more troubling than any particular belief about the authorship of specific Old Testament books that Smith seemed to be advancing. Yet he insisted that Smith's "dogmatism is offensive, his statements are rash, his reasoning inconclusive, his utterances are misleading." What is particularly conspicuous is that Patton directly connected these developments in biblical criticism with evolutionary modes of thinking. "Keunenism"— as he dubbed the historico-critical methods of Abraham Keunen—was nothing other than "the doctrine of evolution applied to the Bible." Indeed, the entire project of the higher critics to reconstruct the authorial history of the Pentateuchal documents was premised on the "hypothesis of evolution." Given that starting point, their conclusions were inevitable. As he explained, "If the doctrine of atheistic evolution is true, if miracles are impossible, if there is no such thing as prophecy . . . then the critics were bound to reach just such conclusions as they have, and it must be stated that they stated their case with great plausibility."[149]

Not surprisingly, in the biological sphere, Patton remained ambivalent about evolution's evidential basis. In keeping with the Princeton scientific and apologetic tradition, he insisted on the central importance of "a teleology immanent in . . . the existing order of biological development." To him, "the doctrine of evolution cannot be made rational without invoking the idea of design . . . the idea of design is woven into the very web of nature." If an anticipatory apologetic is detectable here, even more conspicuous is Patton's choice of rhetoric, which echoed but significantly transformed the words of Charles Hodge: "a more egregious blunder is hardly conceivable than that of assuming that Evolution and Atheism are synonymous." To keep the record straight, he clarified things a few sentences later: "When, therefore, Dr Charles Hodge said of the Darwinian theory that it was Atheism, meaning that it was really a theory of chance, we believe he was right. It is, however, one thing to accept the theory of evolution in general, and another thing to accept a particular theory of evolution."[150]

Promoting anticipatory apologetics was the hallmark of another of Charles Hodge's grandsons, Caspar Wistar Hodge Jr. (1870–1937), whose interventions appropriately serve to complete the circle of Princeton's engagement with Darwin during the era of the Hodge dynasty. He had studied with the eminent Princeton physiological psychologist James Mark Baldwin before undertaking graduate study in Heidelberg and Berlin. In 1907 he secured an assistant professorship in systematic theology, eventually succeeding to Warfield's chair in 1921. As soon as he arrived in the seminary, Warfield assigned him teaching in the anthropology course he had inaugurated in 1888. The approach Wistar Hodge adopted was entirely in keeping with the strategy Warfield had espoused. In the third lecture, he turned to the evolution question, cataloguing the views of such Darwinian detractors such as Emil DuBois-Reymond, Rudolf Virchow, and Hugo de Vries and referring in passing to his cousin William Berryman Scott's work on the evolution of the horse. He also leaned heavily on Henry Fairfield Osborn's account of the tensions between the neo-Darwinians and neo-Lamarckians. It was much the same with the subject of human evolution and the emergence of mind. The absence of paleontological corroboration at the time particularly impressed Hodge, leading him to insist that that theory "is *at least* an unproven doctrine as yet." But this did not prevent him from affirming, in connection with the human body, that "the Christian may allow evolution since the Scripture says it was 'formed' from dust—and it is not necessary to think of this as inorganic clay."

That was the Princeton strategy—hold off on personal judgment in advance of strong empirical confirmation, lay out the evidence warranting caution, and at the same time consider what implications, if any, the theory might hold for theology. In the case of evolution, Hodge's preemptive tactics are crystal clear in his final remarks:

> Even if a genetic connection of the human mind with lower forms of conscious life were proven, this would by no means rule out the category of mediate creation, for in what was specifically new in man, the secondary causes would have to be transcended, and even if active, would not account for the total product. Meanwhile this is a sphere in which all is speculation and where Christian Supernaturalism as to man has nothing to fear from the legitimate results of scientific research. We have thus pointed out the hypothetical character of the evolution theory; and have sought to show that when science did not transgress its limits, no conflict with Christian theology is necessary.[151]

◆ ◆ ◆

THE DEALINGS THAT Presbyterian Princeton had with Darwin's theory during the later nineteenth and early twentieth centuries were a consequence of the unique constellation of forces that prevailed in the seminary and the college during, and immediately after, the McCosh-Patton era. Early on, Princeton was host to two intimately connected but nonetheless distinctive rhetorical spaces sculpted, respectively, by James McCosh and Charles Hodge. At the College of New Jersey, McCosh was passionate about creating a space in which the faith of students would not be jeopardized by attempts to outlaw empirical inquiry or to curb scientific freedom. In that context, he felt obliged to work hard to keep lines of communication open between theological commitment and evolutionary theory and to find ways of cultivating fertile engagements between science and religion. Hodge, by contrast, was long dedicated to the task of maintaining Princeton Seminary as a site of Calvinist conservation, a place for the preservation of Presbyterian Old School values. Deep down, McCosh and Hodge judged Darwinism at the bar of teleology and rejected those forms of evolution that discarded purpose and design, but the rhetorical stances they adopted could not have been more different. For both men, the spaces they occupied shaped their style of speech. Location and locution were intimately connected.

The coming together of these elements produced a distinctive discursive venue at Princeton. In the wake of McCosh's interventions at the college, a succession of prominent scientists, notably William Berryman Scott and Henry Fairfield Osborn, promulgated versions of evolution with a strongly orthogenetic accent.

Their conspicuously non-Darwinian strand of evolutionary thinking showed how it was possible to cast doubts on the supremacy of natural selection while holding to species transformism and descent with modification. That perspective not only allowed their theological colleagues in the seminary to express reservations, with reasonable scientific warrant, about the details of evolutionary mechanisms but also made comprehensible their hesitancy to commit themselves to any particular narrative of evolutionary history. It permitted them to capitalize on the disarray that evolutionary theory found itself in during the decades around 1900 when Darwinism was in eclipse. These circumstances rendered altogether plausible the practice at Princeton Seminary of what I have called anticipatory apologetics, which reiterated that Calvinist orthodoxy had nothing to fear from a broadly evolutionary mindset. For these reasons, Princeton Calvinists found themselves located somewhere between their Presbyterian colleagues in Edinburgh and Toronto, on the one hand, and Belfast and Columbia, on the other. They neither baptized nor bestialized evolutionary theory. As for Darwin himself, they neither vilified nor venerated him. But all the while they reiterated their deep conviction that should evolution come to be verified, it could be Calvinized with little difficulty.[152]

CHAPTER SEVEN

Darwinian Engagements:
Place, Politics, Rhetoric

C HARLES DARWIN had an acute sense of place. Reflecting on what she calls the "special resonance between the man and his domestic setting," Janet Browne insists that without it "Darwin could hardly have hoped to bring his work on natural selection and the origin of species to completion." "Without this sense of place, too," she goes on, "his work would not have taken the singular character it did."[1] Place was personally potent for Darwin, but it also had a critical role to play in his theory of evolution by natural selection. In the *Origin of Species,* he announced that "natural selection will always act according to the nature of the places which are either unoccupied or not perfectly occupied by other beings" and that "natural selection can do nothing . . . until a place in the natural polity of the country can be better filled by some modification of some one or more of its inhabitants."[2] Because adaptation to the particularities of place was the engine power behind evolutionary change, Darwin was persuaded that "varieties would at first generally be local or confined to one place, but if possessed of any decided advantage, or when further modified and improved, they would slowly spread and supplant their parent-forms."[3] Given these professions it is not surprising that Browne subtitles the second volume of her celebrated biography of Darwin "The Power of Place."

If Darwin was a Darwinian through and through, then he could hardly have been surprised by the findings I have presented here. For if the theory of evolution is considered as a sort of conceptual species, then the fact that it fared differently in different places is exactly what Darwin might have predicted. As the theory diffused, it diverged. In different venues, both Darwin's name and Darwinism were made to mean different things. In one place his theory of evolution was seen as an individualist assault on collectivism, in another as a justification for colonial supremacy; elsewhere it was taken to be a subversive attack on racial segregation, yet elsewhere as a symbol of progressive enlightenment. Among the Russian naturalists who undertook scientific exploits in the Siberian wilderness, the Malthusian struggle that Darwin had lodged at the heart of natural selection was downplayed in favor of a cooperative ethic that attributed survival value to the practice of mutual aid. By contrast, the very component of Darwinian evolution that the Saint Petersburg naturalists abominated—struggle—was warmly welcomed by New Zealand evolutionists, who found in it a naturalistic rationale for colonial settlement and Maori dispossession. Among the naturalists who gathered at the Charleston Museum of Natural History in the mid-nineteenth century, it was different yet again. There Darwin's belief that all living beings could be traced back to a common point of origin deeply troubled a scientific community enamored of Agassiz's multiple creations and disturbed to think that the white and black races were of one blood. Darwin's theory was not well adapted to such an environment and so found few advocates. In Cape Town, by contrast, the conversation over Darwin's proposals, at least among English-speaking intellectuals, was conducted with Enlightenment liberality in a setting where the literati aspired to metropolitan values. Racial questions were not prominently on display, perhaps because Darwin's monogenism could easily be made to preserve the unity of the human race while allowing for "improvement" through colonial tutelage and paternalistic governance.

Of course, this list of encounters is selective, not exhaustive, and could easily be extended. What is clear, though, is that in all these venues Darwin's theory signified different things, and its fortunes were conditioned by the environments into which it was introduced. Given these circumstances, it is hardly surprising that in different locations religious communities engaged with Darwin's proposals in markedly different ways. Multiple geographies are at work here. The religious communities on which this analysis concentrates were deeply rooted in Scottish Calvinist culture, but in different places this confessional tradition was marked by the fixations of the society in which it was domesticated. At the same time, because the meaning of Darwin's theory was no less tempered by setting and circumstance, conversations with evolution in different venues displayed markedly distinctive features. Often what looked on the surface like a science-religion fracas, turned out to be about something else—the preservation of cultural identity, the control of educational institutions, sectarian rivalry, race relations. This is not to say, of course, that the Darwinian debates were merely epiphenomena of cultural politics. For one thing, there were concerns that surfaced in many different places—such as the implications of Darwinian biology for teleology, moral sensibility, human nature, and so on. But the deals that interlocutors struck with Darwin bore the stamp of the settings in which they found themselves.

The shape of these encounters with Darwin redraws attention to the salience of *place* and *politics* in religious engagements with scientific claims. But they also point to the role of *rhetoric* in such exchanges, namely, to the significance of style of communication in science-religion debates. The geography of reading thus played a crucial role in intellectual commerce over evolution and in the circuitry of scientific and religious concepts. In different places, texts were read in different ways, and readers distilled different meanings from their rendezvous with words on paper. Why? Because readers bring to their encounter with texts their own reading histories and read in the light of their literary genealogies and cultural preoccupations. The meaning that any new work has for an individual reader is shaped by the other texts, theories, and practices they have already engaged. Meaning bleeds, as it were, from one text to another. New texts take their place within an already established web of textual interlacings. These are manifestly different from person to person, from place to place, from site to site, and they have a key bearing on the spaces of knowledge making. As Gillian Beer tellingly observes, "Books do not stay inside their covers. Once in the head they mingle. The miscegenation of texts is a powerful and uncontrollable force."[4]

The interweaving of place, politics, and rhetoric also surfaced in how people talked about Darwin's speculations, as different spaces conditioned speech in different ways. The relations were reciprocal. The nature of particular spaces was shaped by the style of speech their occupants adopted, and the kind of speech that was permissible was governed by the character of the site. There were marked differences in what could be said and heard in different venues. Subjects that dominated the horizon in one location were absent in another. In one place sectarian politics loomed large; in others, race relations or the heritage of local scientific achievement played a vital role; in yet others it was an enthusiasm for textual criticism or social progressivism. How people spoke about Darwin and dealt with his theory owed much to the preoccupations dominating the conversational culture of local religious communities. When interlocutors resorted to such language as "pigmies in intellect," advocates of "evil," or perpetrators of "bad taste" when depicting opponents—as in Belfast—that marked a discursive tone very different from the tenor of those at Knox College in Toronto who welcomed the "ascent of man" and viewed scripture itself as the product of evolutionary processes.[5] In South Carolina, the characterization of Darwinian science as a mere "pretext for materialism, sensuality, and godlessness," not least because it challenged established racial hierarchy,[6] sounded very different from the view of a prominent Princeton theologian that Darwin was "true, tenderhearted and sympathetic" and "lived a life which has moved the world."[7] Plainly, Darwin's fortunes depended on the theological patois into which his theory was introduced.

Place, Politics and Rhetoric: Then . . .

Our travels around several prominent sites in the global community of Scots Calvinism inspecting how Presbyterians dealt with Darwin have exposed just how superficial monochrome portrayals of the relationship between science and religion as inherently pugilistic or irenic really are. For we have been witness to conflict and conciliation, rejection and endorsement, aggression and appeasement. In different settings, what I call different "flash points" and "trading zones" surfaced.[8] By *flash points* I mean those convictions—cultural, intellectual, doctrinal—that were so central to a community's identity that when challenged by evolutionary thought-forms, anxieties rapidly escalated. In these circumstances, exchanges typically became volatile and debates grew toxic. By *trading zones* I refer to those arenas of engagement where the interface between evolution and theology facilitated fruitful intellectual exchange. The term *trading zone* has been used in anthropological studies to describe something of the processes by which different cultures have been able to exchange commodities despite their differences in language, social relations, and so on. Here I use it to identify arenas in which the use of evolutionary diction encouraged conceptual crossover between scientific exploration and theological inquiry.[9]

In Belfast, the incident that heightened hostilities to evolutionary science was Tyndall's inflammatory presidential speech at the British Association. His naturalistic manifesto, in which he firmly ushered theology off the terrain of natural science, deeply disturbed conservative Calvinists in Ulster's capital. In consequence, they heard nothing but materialism and Epicureanism being promoted with evangelical fervor. In fact, they could hear little else. And with such rhetoric on the lips of one of the scientific establishment's most prominent spokesman, fears for a future state left in the hands of a secularizing scientific intelligentsia were deeply troubling. One axis of tension converged on the management of higher education in a religiously polarized society. The Catholic bishops and archbishops used the opportunity to remind the faithful of their concerns about university education in Ireland and the need to resist any move to deliver pedagogy into the hands of a profane scientific elite. Tyndall's speech only stiffened their resolve to press for a Catholic educational system at all levels. The Protestant leadership in Belfast sought to distance itself from such rhetoric by reminding congregations of the hitherto amicable relations between reformation religion and the cultivation of scientific enterprises. But they worried about the consequences of surrendering all ground to the newly secularized professionals and used the Tyndall episode to insist on the need for theological education. Outside

Belfast, in Derry and Dublin, the cultural politics were different, and Presbyterian voices more sympathetic to Darwin could be heard. Personal pique played its part, too. The refusal of the British Association to allow Robert Watts space to speechify on science and theology infuriated him and added fuel to the fire. The spectacle of Tyndall's public theatrics, a community's sense of being snubbed by the scientific littérateurs, local sectarian politics, and wrangling over the ownership of university education were all flashpoints in Belfast's Calvinist brush with Darwin in the 1870s.

Across the Irish Sea in Edinburgh, by contrast, a productive trading zone had opened up, facilitating rather more fertile dialogue between evolutionary science and dogmatic theology. The long-standing enthusiasm among Free Church intellectuals for scientific enterprises predisposed many in the tradition toward a positive engagement with natural philosophy. At the same time, a growing sense of scripture as a record of progressive revelation encouraged some to find theological sustenance in the idea of society's stadial development. While Belfast Calvinists winced over what they took to be the dark implications of human evolutionary prehistory, key spokesmen for their Scottish coreligionists found little that was objectionable in the idea of humanity's animal ancestry. Indeed, some—like Iverach and Drummond—extracted from the language of cooperative evolution ideas and idioms that could bolster their ethical and political ideologies and lend support to their version of Christianized socialism. Others found in evolution a means of rethinking Paleyite teleology and reenergizing natural theology or of making scientific sense of such human attributes as wonder, awe, and the like. None of this implies that there were no reservations, of course. Critical voices were certainly audible, mostly from those alarmed at the thought of any doctrinal reconstruction or credal refurbishment. Supportive commentators routinely resisted the more aggressive forms of Darwinian imperialism and opposed any inclination toward crude biological reductionism. But taken in the round, the rhetoric was conciliatory and facilitated the cultivation of a transactional space that encouraged intellectual exchange across the borderlands of evolutionary science and evangelical theology.

Naturally, there were flashpoints on the boundaries of scientific scholarship and confessional allegiances. Nowhere did they surface more dramatically than in the heresy hunting of William Robertson Smith, whose mobilization of continental higher criticism and speculative archaeo-anthropology shook the foundations of Scotland's Calvinist orthodoxy. Ironically, while a good deal of Smith's luminous virtuosity actually sprang from an engrained evolutionary progressivism, his radical reconstruction of the compositional history of the Hebrew Bible and his influential genealogy of sacrificial ritual seemed far more menacing than Darwin's theory of species change by means of natural selection. Compared with the abusive rhetoric mounted against Smith and all his works, portrayals of Darwin and his theory were remarkably positive.

Something similar pertained in Toronto, not least at Knox College. A progressivist faith in humanity's gradual spiritual enlargement together with a developmentalist understanding of the religion of the Israelites opened up channels of communication between theological formulation and Darwinian science. The result was that the vocabulary of evolution was happily applied to scripture itself. At the same time, an openness to new thinking on textual criticism fostered an appreciation of the theoretical power of speculation to stimulate fresh lines of inquiry and hence a departure from the strictures of Baconian induction and the sternly practical orientation of scientific endeavor that continued to confine a good deal of natural history throughout Canada. There is also much to be said for the view that Darwinian thought-forms, in one shape or another, fared rather better among the theologians than within the community of natural science practitioners precisely because of this willingness to welcome conjecture and inference. Still, for all the Baconian veneration, there were definite moments when Canadian science rose above its faith in mere facticity. From his position as president of the University of Toronto, Daniel Wilson worked hard to get Toronto on the empire's map of metropolitan science, welcomed Alfred Russel Wallace to the campus, and spoke cordially of Darwin's theory even if he could not go the whole way with him. Nothing occurred in Toronto like the public spectacle of either Tyndall's Belfast cannonbolt or the Smith heresy hunt in the Scottish Free Kirk to incline interlocutors in one direction or another. Again, this does not mean that dissenting voices were entirely absent. From his powerhouse at McGill in Montreal, for example, John William Dawson issued periodic critique, though that was never sufficient to encourage him to join forces with Quebec's Catholic anti-Darwinians. Still, taken overall, the tone of the Darwinian conversation, particularly among Scots Calvinists, in Toronto was cordial and conspicuously bereft of the aggression discernible among their religious counterparts elsewhere.

Things were very different in Columbia, South Carolina—not least for James Woodrow, who lost his position at the seminary over his evolutionary views. A profound aversion to any modernizing impulses, born in large part of the South's struggle with northern abolitionism, made it suspicious of successive waves of ever-newer scientific declarations. To figures like Robert Dabney, science did little more than give him one headache after another. Geologists had thrown doubts on biblical cosmology and chronology, anthropologists had sacrificed the unity of the human race on the altar of scientific polygeny, and now evolutionary theory was undermining species integrity and human identity. The rigid boundaries and essential hierarchies that the creator had lodged at the heart of nature and society alike were facing meltdown as science whittled away at the foundations of the divine order and replaced fixity with fluidity. In these circumstances, any compromise with Darwin was seen as perfidy. Besides, there were pressing political reasons to remain loyal to a literal bible. As southern theologians did not weary of pointing out, a plain, straightforward, no-nonsense reading of the Old and New Testaments had provided perfectly clear warrant for a slave system. Even after the Civil War, a yearning for the lost cause of the Confederacy made Southern Presbyterians resistant to any tinkering with the Mosaic narrative to suit the whims of the latest scientific fashion. Fancy hermeneutics had done untold damage in the North, and those who toyed with metaphor, myth, and symbol in reading Genesis were thundering headlong toward spiritual apostasy and political chaos. In that environment, it was extraordinarily difficult to separate out scientific findings, materialist philosophy, social reconstruction, and cultural politics. For such reasons, discussions over evolution had more than their fair share of talk about "infidelity," "sensuality," "wild speculation," "vanity," and "deceit." In such a rhetorical space the possibilities for any productive engagement with evolution were slim, and those who tried to make deals with Darwin were rewarded with charges of heresy, treachery, and betrayal.

At American Calvinism's nerve center—Princeton, New Jersey—the tone was different again. Several things conspired here to produce a distinctive style of dialogue over Darwin. At the College of New Jersey, there was the towering presence of James McCosh, whose temperate tones on the evolution question kept the institution open to biological theorizing and cultivated the creativity of a number of leading evolutionists in the era. At Princeton Seminary, Charles Hodge's diagnosis of Darwinism as atheism sounded very different, though on closer inspection he and McCosh shared the same concerns over the challenge to teleology. In some part at least, the language—the rhetoric—they chose to express their views on the whole subject reflected the politics of their location in the two different Princeton establishments—college and seminary. At the same time, George Macloskie pushed evolution's virtues to readers of the Princeton journals, while B. B. Warfield kept an open mind on the question. Macloskie, who taught biology at the college, was well placed to provide informed commentary on evolution's scientific credibility; Warfield's interest sprang in part from extensive firsthand experience of cattle breeding on his father's stock farm, which kept him in close touch with scientific thinking on such key subjects as heredity, variation, reversion to type, and other matters central to evolutionary theory. These factors, together with the traditionalist conception of biblical inspiration that circulated at the seminary, made the Princeton theologians cautious in their judgments about Darwinism narrowly construed. In adopting this stance, they enjoyed the support of a number of leading biologists and paleontologists at the university who culti-vated distinctive forms of non-Darwinian evolution. Their collective judgment displayed neither the scornful contempt of Columbia or Belfast, nor the warmer embrace of Edinburgh or Toronto. Their endorsements were moderate, cautious, guarded, but that did not prevent one leading spokesman from insisting that Calvin's doctrine of the creation was evolutionism of an extremely pure variety. Of one thing the Princetonians were sure: if evolutionary theory were to be veri-fied, Calvinist theology could easily ride out the storm.

Place, Politics and Rhetoric: . . . and Now

The sites we have visited in our scrutiny of how a suite of related communities dealt with Darwin in the decades around 1900 disclose the power of space and place, cultural politics and rhetorical style in shaping dialogues over scientific questions. I suspect it is still much the same. And so, as an exploratory postscript from the vantage point of our own time, the following reflections on a couple of contemporary feuds over Darwin are intended to gesture toward something of how place, politics, and rhetoric may continue to govern what might be called the history of the Darwinian present.[10]

It's the morning of Friday, 2 July 2010. The place is Lecture Theatre 1.31, Im-perial College London. Keith Bennett, a paleoecologist, is there to deliver a keynote speech to the International Paleontological Congress in the session "Macro-evolution and the Modern Synthesis."[11] Bennett's aim is to assess the role of adaptation to environmental change in evolutionary history, and it so captures the imagination of Graham Lawton, commissioning editor for the *New Scientist,* that he invites the speaker to publish his thoughts in the magazine.

The article would turn out to make the front cover of the journal. Bennett's conclusion was clear. "Major climatic events such as ice ages," he observed, "ought to leave their imprint on life as species adapt to new conditions." Was that the case? Bennett didn't think so. He suspected that pan-adaptationism might turn out not to be the major driver of evolutionary change that it has routinely been assumed to be. As he put it, "the connection between environmental change and evolutionary change is weak, which is not what might have been expected from Darwin's hypothesis."[12] In its place, he suggested that evolution may be nonlinear and that the causes of macroevolutionary change lie in the "chaotic" dynamics of the relationship between genotype and phenotype.

The piece was quickly picked up in what may be called the blogosphere, particularly the blog run by the Chicago evolutionist Jerry Coyne. He verged on the apoplectic, as his rhetoric amply disclosed. In a nutshell, Coyne thought Bennett's piece "stupid," "thoughtless," "hogwash," "drivel," "ludicrous," "ignorant."[13] On the day he posted his assessment, forty-two responses were logged. One worried that "it would be misinterpreted in the wrong corner"; another found it "sad and pathetic"; yet another speculated that Bennett had been "bitten by a Postmodernist" and warned that "those bites can get infected"! A day or two later, following another injection from Coyne, a further sixty comments registered a renewed adrenalin rush of scorn. Many complained about the follies of science journalism; some claimed, missing the irony that blogs themselves are not subject to peer review, that Bennett's ideas wouldn't get a hearing "in a 'real' science journal" that uses proper expert evaluation.

The contrast between Bennett's reception in Coyne blog-land and at Imperial College could not be sharper. As Lawton observed in a message to Coyne, "if Bennett is so hopelessly wrong, why was he ever invited to give that keynote (alongside Niles Eldredge)? Why did the symposium even take place? Bennett was not the only one to question the primacy of natural selection in macroevolution. Why does the Royal Society support his work? . . . I was at Bennett's talk; the room was full of learned and eminent people. He took a few questions but there were no howls of protest like yours. What am I to make of this?"[14] Besides, the substance of Bennett's argument had been presented a few years earlier at the Royal Society and published—in a fully refereed journal, it should be noted—in its *Philosophical Transactions*.[15] Again, that audience engaged calmly with his proposals and displayed none of the invective that dramatically surfaced in the blogosphere.

The different fate of Bennett's analysis in these different arenas reinforces the significance of different speech spaces in scientific disputes. What can be said and—no less important—what can be heard in different venues is critical here. Plainly, the expert audiences that gathered at Imperial College London in 2010 and at the Royal Society in 2003 heard neither the idiocy nor the illogicality of the kind that Coyne and company found written all over the published version that appeared in a popular science journal. Obviously, the different settings in which Bennett's ideas featured conditioned what interlocutors heard and what they said. Place, poetics and polemics were intimately connected.

Bennett's story is not unique. But if it displays the salience of place and rhetoric in the Darwin debates, Jerry Fodor's experience discloses the significance of cultural politics. Fodor is State of New Jersey professor of philosophy at Rutgers University and a world-leading philosopher of mind. His querying of Darwinian adaptationism, initially in the context of evolutionary psychology, likewise drew venomous commentary from the ranks of the High Darwinians. The details of his argument need not detain us here, save to say that Fodor argued that in his use of the metaphor of natural selection, Darwin did not succeed in getting mind out of nature. This conclusion didn't mean that Fodor rejected "the central Darwinist theses of the common origin and mutability of species"; it was just that he was now convinced that the standard pan-adaptationist explanation of evolutionary change by natural selection could not do the comprehensive job required of it.[16]

What really propelled him into the public limelight, however, was the book he coauthored with Massimo Piattelli-Palmarini in 2010, entitled *What Darwin Got Wrong*. It attracted a blizzard of protest. A mere sampling of the vilification suffices to give a flavor of the whole. One article was titled "Two Critics Without a Clue"; others portrayed Fodor and Piattelli-Palmarini as "sterile and wrong-headed," "willfully ignorant," "arrogant and obfuscating," and "dangerous" and described their book as "predictably ripe for making mischief."[17] Why all the malice?

To nineteenth-century Calvinists in the American South, Darwin's name triggered horror about the fluidity of species and its implications for race relations; in Victorian Belfast, his science sounded the alarm about the control of education in a sectarian society. In the early twenty-first century, *any* querying of Darwinism conjures up the specter of such bêtes noires as creationism and intelligent design. To get properly on track here, we need to record Fodor and Piattelli-Palmarini's views on religion. They confess themselves "card-carrying, signed-up, dyed-in-the-wool, no-holds-barred atheists."[18] Despite that announcement, reviewers persistently cast their text into the arena of science-and-religion debate. Dennett had already declared that people like Fodor who are not convinced that natural selection can successfully ground a theory of natural teleology need to be properly labeled: "we call them creationists," he declared.[19] Robert Richards told readers of the *American Scientist* that the authors had orchestrated "a medley of contradictions that can delight only the ears of creationists and proponents of intelligent design" and concluded his review by citing—presumably as a parable—the Wilberforce-Huxley legend.[20] Even those reviewers who largely concentrated their energies on the book's arguments included throw-away allusions to fundamentalist obscurantism. Michael Ruse was perhaps the most forthright. Writing in the *Boston Globe,* he concluded, "Like those scorned Christians, Fodor and Piattelli-Palmarini just cannot stomach the idea that humans might just be organisms, no better than the rest of the living world . . . Christians are open in their beliefs that humans are special and explaining them lies beyond the scope of science. I just wish that our authors were a little more open that this is their view too."[21] Naturally enough, Fodor and Piattelli-Palmarini didn't take too kindly to that. After all, the very first two sentences of their book ran: "This is not a book about God; nor about intelligent design; nor about creationism. Neither of us is into any of those."[22]

Plainly, whatever the rightness or wrongness of Keith Bennett's doubts about the role of adaptationism in evolutionary history or of Jerry Fodor's assault on the coherence of natural selection, the uproar their work generated shows that Darwin's name remains as iconic now as it was in the nineteenth century and that what it signifies is different in different communities. If my suspicions are well founded, I believe it also shows just how pervasive—in one way or another—place, politics, and rhetoric continue to be in dealing with Darwin.

A bibliography is available at the publisher's website (www.jhu.press.edu); search for this book by title or by author for a link to the bibliography.

Chapter 1 · Dealing with Darwin

1. Edward Said, "Traveling Theory," reprinted as chapter 10 of *The World, the Text and the Critic* (London: Vintage, 1991), p. 226.

2. See James A. Secord, "Knowledge in Transit," *Isis* 95 (2004): 654–672, on pp. 663–664.

3. See David N. Livingstone, "Science, Text and Space: Thoughts on the Geography of Reading," *Transactions of the Institute of British Geographers* 35 (2005): 391–401; and David N. Livingstone, "Science, Site and Speech: Scientific Knowledge and the Spaces of Rhetoric," *History of the Human Sciences* 20 (2007): 71–98. The following paragraphs are drawn from these pieces.

4. Nicolaas Rupke, "A Geography of Enlightenment: The Critical Reception of Alexander von Humboldt's Mexico work," in David N. Livingstone and Charles W. J. Withers, eds., *Geography and Enlightenment* (Chicago: University of Chicago Press, 1999), 281–294, on p. 336.

5. James A. Secord, *Victorian Sensation: The Extraordinary Publication, Reception, and Secret Authorship of Vestiges of the Natural History of Creation* (Chicago: University of Chicago Press, 2001), pp. 14, 24.

6. Diarmid A. Finnegan, "The Work of Ice: Glacial Theory and Scientific Culture in Early Victorian Edinburgh," *British Journal for the History of Science* 37 (2004): 29–52.

7. See Steven Shapin, *The Scientific Revolution* (Chicago: University of Chicago Press, 1996), and Charles W. J. Withers, *Placing the Enlightenment: Thinking Geographically about the Age of Reason* (Chicago: University of Chicago Press, 2007).

8. Rebekah Higgitt, *Recreating Newton: Newtonian Biography and the Making of Nineteenth-Century History of Science* (London: Pickering & Chatto, 2007).

9. David N. Livingstone, "Politics, Culture and Human Origins: Geographies of Reading and Reputation in Nineteenth Century Science," in David N. Livingstone and Charles W. J. Withers, eds., *Geographies of Nineteenth Century Science* (Chicago: University of Chicago Press, 2011), pp. 178–202.

10. Nicolaas A. Rupke, *Alexander von Humboldt: A Metabiography* (Chicago: University of Chicago Press, 2008), p. 16. For a similar study of the changing posthumous reputations of the missionary explorer David Livingstone, see Justin D. Livingstone, *Livingstone's 'Lives': A Metabiography of a Victorian Icon* (Manchester: Manchester University Press, 2014).

11. Owen Gingerich, *The Book Nobody Read: Chasing the Revolutions of Nicolaus Copernicus* (London: Heinemann, 2004), p. 255.

12. Innes M. Keighren, "Bringing Geography to the Book: Charting the Reception of *Influences of Geographic Environment,*" *Transactions of the Institute of British Geographers* 31 (2006): 525–540. See also Innes M. Keighren, *Bringing Geography to Book: Ellen Semple and the Reception of Geographical Knowledge* (London: I. B. Tauris, 2010).

13. Bruno Latour, *Science in Action* (Milton Keynes: Open University Press, 1987).

14. Marwa Elshakry, "Knowledge in Motion: the Cultural Politics of Modern Science Translations in Arabic," *Isis* 99 (2008): 701–730 on pp. 702, 703.

15. James Clifford, *The Predicament of Culture* (Cambridge: Cambridge University Press, 1988).

16. Stanley Fish, *Is There a Text in this Class? The Authority of Interpretive Communities* (Cambridge, MA: Harvard University Press, 1980).

17. Jonathan Rose, "Arriving at a History of Reading," *Historically Speaking* 5, no. 3 (2004): 36–39, on p. 39.

18. Theodore Zeldin, *Conversation* (London: Harvill Press, 1998), pp. 14, 7. See also Stephen Miller, *Conversation: A History of a Declining Art* (New Haven: Yale University Press, 2006).

19. Quoted in Steven Shapin, "Cordelia's Love: Credibility and the Social Studies of Science," *Perspectives on Science* 3 (1995): 255–275, on p. 256.

20. See Alice Walters, "Conversation Pieces: Science and Polite Society in Eighteenth-Century England," *History of Science* 35 (1997): 121–154.

21. Joseph Addison, *Essays Moral and Humorous and Essays on Imagination and Taste* (Edinburgh: Chambers, 1839), p. 177.

22. For geographical perspectives on such matters, see Miles Ogborn, "The Power of Speech: Orality, Oaths and Evidence in the British Atlantic World, 1650–1800," *Transactions of the Institute of British Geographers* 36 (2011): 109–125; Miles Ogborn, "Talking Plants: Botany and Speech in Eighteenth-Century Jamaica," *History of Science* 51 (2013): 251–282.

23. See, e.g., Geoffrey V. Sutton, *Science for a Polite Society: Gender, Culture, and the Demonstration of Enlightenment* (Boulder: Westview, 1995); Mary Terrall, "Salon, Academy, and Boudoir: Generation and Desire in Maupertuis's Science of Life," *Isis* 87 (1996): 217–229.

24. James A. Secord, "How Scientific Conversation Became Shop Talk," in Aileen Fyfe and Bernard Lightman, eds., *Science in the Marketplace* (Chicago: University of Chicago Press, 2007), pp. 23–59.

25. Peter Burke, *Languages and Communities in Early Modern Europe* (Cambridge: Cambridge University Press, 2004), p. 6.

26. James Scott, *Weapons of the Weak: Everyday Forms of Peasant Resistance* (New Haven: Yale University Press, 1985).

27. John Hedley Brooke, "Response," *Historically Speaking* 8, no. 7 (May/June 2007), 16–17.

28. Joseph Hooker to Charles Darwin, 6 Oct. 1865, in Frederick H. Burkhardt, ed., *The Correspondence of Charles Darwin,* vol. 13 (Cambridge: Cambridge University Press, 2002), pp. 261.

29. John Hedley Brooke, "The Relations Between Darwin's Science and his Religion," in John Durant, ed., *Darwinism and Divinity: Essays on Evolution and Religious Belief* (Oxford: Basil Blackwell, 1985), pp. 40–75, on p. 40.

30. Matthew Day, "Godless Savages and Superstitious Dogs: Charles Darwin, Imperial

Ethnography and the Problem of Human Uniqueness," *Journal of the History of Ideas* 69 (2008): 49–70, on p. 55.

31. Winchell's letter is reproduced in L. Alberstadt, "Alexander Winchell's Preadamites— A Case for Dismissal from the Vanderbilt University," *Earth Sciences History* 13 (1994): 97–112.

32. J. R. Lucas, "Wilberforce and Huxley: A Legendary Encounter," *Historical Journal* 22 (1979): 313–330, on p. 327.

33. I have argued for the significance of place for understanding scientific enterprises in David N. Livingstone, *Putting Science in Its Place: Geographies of Scientific Knowledge* (Chicago: University of Chicago Press, 2003).

34. Thomas F. Glick, "Preface, 1988: Reception Studies Since 1974," in Thomas F. Glick, ed., *The Comparative Reception of Darwinism* (Chicago: University of Chicago Press, 1988), pp. xi–xxviii, on p. xxviii. David Hull, "Evolutionary Thinking Observed," *Science* 223 (1984): 923–924, on p. 923.

35. Eve-Marie Engels and Thomas F. Glick, "Editors' Introduction," in Eve-Marie Engels and Thomas F. Glick, eds., *The Reception of Charles Darwin in Europe*, vol. 1 (London: Continuum, 2008), pp. 1–22, on pp. 1, 2, 7.

36. Marwa Elshakry, "Global Darwin: Eastern Enchantment," *Nature* 461 (29 Oct. 2009): 1200–1201, on p. 1201.

37. James Pusey, "Global Darwin: Revolutionary Road," *Nature* 462 (12 Nov. 2009): 162–163, on p. 162.

38. Jürgen Buchenau, "Global Darwin: Multicultural Mergers," *Nature* 462 (19 Nov. 2009): 284–285, on p. 284.

39. See Lester D. Stephens and Dale R. Calder, "John McCrady of South Carolina: Pioneer Student of North American Hydrozoa," *Archives of Natural History* 19 (1992): 39–45.

40. John McCrady, "The Law of Development by Specialization: A Sketch of Its Probable Universality," *Journal of the Elliott Society* 1 (1860): 101–114, on p. 102.

41. John McCrady, "The Study of Nature and the Arts of Civilized Life," *De Bow's Review* 30 (1861): 579–606, on p. 604.

42. McCrady, "Study of Nature," pp. 595, 597.

43. See Edward Lurie, "Louis Agassiz and the Races of Man," *Isis* 45 (1954): 227–242; Mary P. Winsor, "Louis Agassiz and the Species Question," in William Coleman and Camille Limoges, eds., *Studies in History of Biology* (Baltimore: Johns Hopkins University Press, 1979), pp. 89–117.

44. What follows draws on Lester D. Stephens, *Science, Race, and Religion in the American South: John Bachman and the Charleston Circle of Naturalists, 1815–1895* (Chapel Hill: University of North Carolina Press, 2000).

45. This does not mean that all the Charleston naturalists without exception read evolution in precisely the same way nor that opponents always shared the same grounds for their disquiet. John Bachman, for example, long believed in the permanence of species and rejected Lamarck's "absurd" notions about how species were supposed to undergo transmutation. But he repudiated the way Agassiz mobilized ideas about natural provinces and centers of creation to justify his belief in the existence of different human species. And yet he too retained a life-long commitment to the idea that certain races were stamped with inferiority. See John Bachman, *The Doctrine of the Unity of the Human Race Examined on the Principles of Science* (Charleston: Canning, 1850).

46. The plurality of opinions on Darwinism in the South more generally is stressed in Ronald L. Numbers and Lester D. Stephens, "Darwinism in the American South," in Ronald L. Numbers and John Stenhouse, eds., *Disseminating Darwinism: The Role of Place, Race, Religion, and Gender* (New York: Cambridge University Press, 1999), pp. 123–143.

47. On the reaction of New Zealanders to evolution, I have benefited greatly from John Stenhouse, "The Darwinian Enlightenment and New Zealand Politics," in Roy MacLeod and Philip F. Rehbock, eds., *Darwin's Laboratory: Evolutionary Theory and Natural History in the Pacific* (Honolulu: University of Hawai'i Press, 1994), pp. 395–425, and John Stenhouse, "Darwinism in New Zealand, 1859–1900," in Numbers and Stenhouse, *Disseminating Darwinism*, pp. 61–89.

48. William T. L. Travers, "On the Changes Effected in the Natural Features of a New Country by the Introduction of Civilized Races," *Transactions and Proceedings of the New Zealand Institute* 2 (1869): 299–313, on p. 313, 308, 313.

49. Alfred K. Newman, "A Study of the Causes Leading to the Extinction of the Maori," *Transactions and Proceedings of the New Zealand Institute* 14 (1881): 459–477, on pp. 475, 470. On Newman, see John Stenhouse, "'A Disappearing Race before We Came Here': Doctor Alfred Kingcome Newman, the Dying Maori, and Victorian Scientific Racism," *New Zealand Journal of History* 30 (1996): 124–140.

50. Walter L. Buller, "Presidential Address to the Wellington Philosophical Society," *Transactions and Proceedings of the New Zealand Institute* 17 (1884): 443–446, on pp. 443, 444. For Buller's application of natural selection to ornithological questions, see Walter Buller, "Illustrations of Darwinism; or, The Avifauna of New Zealand Considered in Relation to the Fundamental Law of Descent with Modification," *Transactions and Proceedings of the New Zealand Institute* 27 (1894): 75–104.

51. Stenhouse, "Darwinism in New Zealand," p. 81.

52. Charles Darwin, *The Descent of Man and Selection in Relation to Sex*, 2nd ed. (London: John Murray, 1871), p. 128.

53. Joseph Giles, "Waitara and the Native Question," *Southern Monthly Magazine* 1 (1863): 209–216, on p. 215.

54. Quotations are to be found in John Stenhouse, "The Evolution Debates in Nineteenth-Century Dunedin," unpublished manuscript; and also in Margaret Tennant, "MacGregor, Duncan. 1843–1906," *Dictionary of New Zealand Biography*, updated 22 June 2007, and available at www.dnzb.govt.nz/DNZB/alt_essayBody.asp?essayID=2M7. I am most grateful to Dr. Stenhouse for allowing me to see his unpublished analyses.

55. Quoted in Daniel P. Todes, *Darwin Without Malthus: The Struggle for Existence in Russian Evolutionary Thought* (Oxford: Oxford University Press, 1989), pp. 110, 111.

56. See Todes, *Darwin Without Malthus*, chapter 6.

57. On Kropotkin, see George Woodcock and Ivan Avakumovic, *The Anarchist Prince: A Biographical Study of Peter Kropotkin* (London: T. V. Boardman, 1950), and Martin A. Miller, *Kropotkin* (Chicago: University of Chicago Press, 1976).

58. Peter Kropotkin, *Mutual Aid: A Factor of Evolution* (Harmondsworth: Penguin, 1939, first pub. 1902), pp. 26–27.

59. Quoted in Daniel P. Todes, "Darwin's Malthusian Metaphor and Russian Evolutionary Thought, 1859–1917," *Isis* 78 (1987): 537–551, on p. 546.

60. See Todes, *Darwin Without Malthus*, chapter 2.

61. James Allen Rogers, "Russian Opposition to Darwinism in the Nineteenth Century," *Isis* 65 (1974): 487–505, on p. 497.

62. See George L. Kline, "Darwinism and the Russian Orthodox Church," in Ernest J. Simmons, ed., *Continuity and Change in Russian and Soviet Thought* (Cambridge, MA: Harvard University Press, 1955), pp. 307–328.

63. This is not to say that Darwinian vocabulary was never deployed for racial purposes, particularly in later decades, in southern Africa. George W. Stow resorted to Darwinian talk of a struggle for existence in his account of the Bushmen, and later George McCall Theal found the idea of the survival of the fittest suited his convictions about the triumph of white settlement. See Saul Dubow, *Scientific Racism in Modern South Africa* (Cambridge: Cambridge University Press, 1995). For a fuller account of the fortunes of Darwin in Cape Town, see David N. Livingstone, "Debating Darwin at the Cape," in Patrick Harries and Martin Lingwiler, eds., "Knowing Worlds: Science between Africa and Europe," unpublished manuscript.

64. Saul Dubow, "Earth History, Natural History, and Prehistory at the Cape, 1860–1785," *Comparative Studies in Society and History* 46 (2004): 107–133, on p. 109.

65. Saul Dubow, *A Commonwealth of Knowledge: Science, Sensibility, and White South Africa 1820–2000* (Oxford: Oxford University Press, 2006), p. 71.

66. J. I. Jitse Van Rensburg, "Dale, Sir Langham," in W. J. De Kock, ed., *Dictionary of South African Biography*, vol. 1 (1968), pp. 201–204.

67. Langham Dale, "Anthropology—A Review of Modern Theories," *Cape Monthly Magazine* 8 (June 1874): 350–362, on pp. 355, 356, 357.

68. Langham Dale, "On the Origin of Language," *Cape Monthly Magazine* 9 (July 1874): 15–19, on p. 18.

69. Biographical details are available in T.R.H.D., "Berry, Sir William Bisset," in W. J. De Kock and D. W. Krüger, eds., *Dictionary of South African Biography*, vol. 2 (1972), pp. 55–56; see also J. K. De Kock, "Doctors in Parliament," *South African Medical Journal* 38 (11 Apr. 1964): 237–241.

70. W. B. B[erry], "Dr. Dale on Evolution of Life. A Criticism in Two Chapters. Chap. 1," *Cape Monthly Magazine* 9 (Aug. 1874): 95–106; W. B. B[erry], "Dr. Dale on Evolution of Life. A Criticism in Two Chapters. Chap. 2," *Cape Monthly Magazine* 9 (Sept. 1874): 151–158, on pp. 157, 158.

71. Langham Dale, "The Library Address," *Cape Monthly Magazine* 9 (Oct. 1874): 219–222, on p. 222.

72. Biographical details are available in J. L. McCracken, *New Light at the Cape of Good Hope: William Porter, the Father of Cape Liberalism* (Belfast: Ulster Historical Foundation, 1993).

73. Indeed, his brother had put pen to paper in October of that year to counter Tyndall's account of atomism and the several species of atheism with which it was associated. J. Scott Porter, *The Existence and Attributes of the Invisible God Made Manifest through his Visible Works: A Discourse Delivered on the 30th August, 1874 in the First Presbyterian Church, Rosemary Street, Belfast; In Which Some Statements Made by Professor Tyndall in his Inaugural Address, Read at the Annual Meeting of the British Association for the Advancement of Science, are Briefly Considered* (Belfast: Wm. Henry Greer, 1874).

74. William Porter, "On Scholastic and Philosophic Studies," *Cape Monthly Magazine* 13 (Aug. 1876): 65–81, on pp. 77, 78.

75. The best general guide to the history of relations between science and religion is John Hedley Brooke, *Science and Religion: Some Historical Perspectives* (Cambridge: Cambridge University Press, 1991).

76. Andrew Dickson White, "New Chapters in the Warfare of Science. XIX—From Creation to Evolution," *Popular Science Monthly* 45 (1894): 145–159, on p. 157.

77. I have discussed this episode in *Adam's Ancestors: Race, Religion and the Politics of Human Origins* (Baltimore: Johns Hopkins University Press, 2008).

78. Andrew Dickson White, *A History of the Warfare of Science with Theology in Christendom*, 2 vols. (New York: Appleton, 1898), 1:240–242.

79. Cited in Alberstadt, "Alexander Winchell's Preadamites," p. 108.

80. See William Stanton, *The Leopard's Spots: Scientific Attitudes toward Race in America, 1815–59* (Chicago: University of Chicago Press, 1960); Thomas F. Gossett, *Race: The History of an Idea in America* (Dallas: Southern Methodist University Press, 1963).

81. Alexander Winchell, "Preadamite," in John McClintock and James Strong, eds., *Cyclopaedia of Biblical, Theological, and Ecclesiastical Literature*, vol. 8 (New York: Harper and Brothers, 1877), p. 484.

82. Alexander Winchell, *Adamites and Preadamites: or, a Popular Discussion Concerning the Remote Representatives of the Human Species and their Relation to the Biblical Adam* (Syracuse: John T. Roberts, 1878), p. 24.

83. Cited in Mark David Wood, "Debating Science and Religion: Towards a Comparative Geography of Public Controversy, 1874–1895" (Ph.D. thesis, Queen's University Belfast, 2011), p. 147.

84. Cited in Alberstadt, "Alexander Winchell's Preadamites," p. 110.

85. Niels Henrik Gregersen and Peter C. Kjærgaard, "Darwin and the Divine Experiment: Religious Responses to Darwin in Denmark, 1859–1909," *Studia Theologica* 63 (2009): 140–161, on p. 141.

86. Hans Henrik Hjermitslev, "Protestant Responses to Darwinism in Denmark, 1859–1914," *Journal of the History of Ideas* 72 (2011): 279–303, on pp. 281, 303.

87. The following couple of paragraphs rely on the work of my former Ph.D. student, Mark David Wood, "Debating Science and Religion."

88. Thomas MacLaughlin, *Gaelic Gleanings; or, Notices of the History and Literature of the Scottish Gael* (Edinburgh: MacLaughlin and Stewart, 1862), p. 166.

89. Patrick Carnegie Simpson, *The Life of Principal Rainy*, 2 vols. (London: Hodder and Stoughton, n.d., ca. 1909), 1:429, 430.

90. Quoted in Wood, "Debating Science and Religion," p. 193, from "Rev. Mr Macaskill on Evolution: Denunciation of Professor Drummond's Book: Discussion by Dingwall Presbytery," *Ross-shire Journal* (15 Feb. 1895).

91. White, *History of the Warfare*, p. 84.

92. What follows draws on Marwa Elshakry, "The Gospel of Science and American Evangelism in Late Ottoman Beirut," *Past and Present* 196 (2007): 173–214.

93. Samir Kassir, *Beirut*, translated by M. B. Debevoise (Berkeley: University of California Press, 2010), pp. 187–188.

94. Elshakry, "The Gospel of Science," p. 213. See also her more detailed analysis of al-Jisr in Marwa Elshakry, "Muslim Hermeneutics and Arabic Views of Evolution," *Zygon* 46 (2011): 330–344.

Chapter 2 · Edinburgh, Evolution, and Cannibalistic Nostalgia

1. See H. C. G. Matthew, "Rainy, Robert (1826–1906)," *Oxford Dictionary of National Biography* (Oxford: Oxford University Press, 2004).

2. Robert Rainy, *Evolution and Theology: Inaugural Address* (Edinburgh: Maclaren & Macniven, 1874), pp. 7, 9, 11, 32.

3. Rainy, *Evolution and Theology*, pp. 32, 17.

4. See John Duns, "The Origin of Species," *North British Review* 32 (1860): 455–486.

5. John Duns, *Biblical Natural Science, Being The Explanation of all References in Holy Scripture to Geology, Botany, Zoology, and Physical Geography*, vol. 1 (London: William Mackenzie, 1863), pp. 568, 539.

6. John Duns, *Science and Christian Thought* (London: Religious Tract Society, [circa 1866]), pp. 207, 301, 292.

7. John Duns, *Creation According to the Book of Genesis and the Confession of Faith. Speculative Natural Science and Theology. Two Lectures* (Edinburgh: Maclaren & Macnivan, 1877), p. 36.

8. John Duns, *On the Theory of Natural Selection and the Theory of Design: Being a Paper Read before the Victoria Institute, or Philosophical Society of Great Britain* (London: Victoria Institute, 1887). Duns also maintained this viewpoint in *Christianity and Science* (Edinburgh: William P. Kennedy, 1860), and *Science and Christian Thought*.

9. David Brewster, "The Facts and Fancies of Mr Darwin," *Good Words* 3 (1862): 3–9, on pp. 3, 8.

10. Free Church responses appeared in "The Philosophical Institution and Professor Huxley," *Witness*, 14 Jan. 1862, and "Professor Huxley's Theory of the Origin and Kindred of Man," *Witness*, 18 Jan. 1862.

11. For both an account of this event and the source of this quotation, see Adrian Desmond, *Huxley: The Devil's Disciple* (London: Michael Joseph, 1994), pp. 300–301.

12. The lecture was advertised in the *Caledonian Mercury*, 14 Oct. 1861, and the proceeds were donated to the Royal Infirmary. Details of the lecture were reported in *The Caledonian Mercury*, 19 Oct. 1861. Paul B. du Chaillu was the author of a work on African exploration, *Explorations and Adventures in Equatorial Africa: With Accounts of the Manners and Customs of the People, and of the Chace of the Gorilla, Crocodile, Leopard, Hippopotamus, and Other Animals* (London: John Murray, 1861).

13. Reported in "Glasgow Night Asylum for the Houseless—M. Du Chaillu," *Scotsman*, 8 Oct. 1861.

14. James Scowen, "A Study in the Historical Geography of an Idea: Darwinism in Edinburgh, 1859–75," *Scottish Geographical Journal* 114 (1998): 148–156. On du Chaillu more generally, see Stuart McCook, "'It May Be Truth, but It Is Not Evidence': Paul du Chaillu and the Legitimation of Evidence in the Field Sciences," *Osiris* 11 (1996): 177–197; K. David Patterson, "Paul du Chaillu and the Exploration of Gabon, 1855–1865," *International Journal of African Historical Studies* 7 (1974): 647–667.

15. Thomas Smith, *Natural Laws: Creation, Miracles, Providence, Prayer, A Lecture Delivered at the Request of the Young Men's Association in Connexion with the U.P. Congregation, Broughton Place, Edinburgh* (Edinburgh: Duncan Grant, 1867), pp. 16, 15. In a comparable lecture he delivered the previous year, Smith drew attention to Darwin's efforts "to prove

the generic identity of man with the higher order of monkeys." His response was that "there is not a single fact on which the hypothesis rests." Thomas Smith, "The Bible not Inconsistent with Science," in *Christianity and Recent Speculations. Six Lectures by Ministers of the Free Church with a Preface by Robert S. Candlish* (Edinburgh: John Maclaren, 1886), pp. 1–32, on p. 28.

16. See Crosbie Smith and M. Norton Wise, *Energy and Empire: A Biographical Study of Lord Kelvin* (Cambridge: Cambridge University Press, 1989); Joe D. Burchfield, *Lord Kelvin and the Age of the Earth* (Chicago: University of Chicago Press, 1975).

17. J. D. Forbes, "On the Antiquity of Man," *Good Words* 5 (1864): 253–258, 432–440, on p. 440.

18. David Guthrie, *Christianity and Natural Science: A Lecture Delivered Before the Young Men's Association of the Free Tolbooth Church* (Edinburgh, 1866).

19. John Tulloch, "Modern Scientific Naturalism," *Blackwood's Magazine* 116 (1874): 519–539.

20. Robert Flint, *Theism* (Edinburgh: Blackwood & Sons, 1877), pp. 201, 202, 203, 208, 205, 198, 205, 198.

21. Robert Flint, *Anti-theistic Theories* (Edinburgh: Blackwood & Sons, 1879), p. 100.

22. Flint, *Theism*, pp. 208–209.

23. Biographical details are available in W. L. Calderwood and D. Woodside, *The Life of Henry Calderwood* (London: Hodder and Stoughton, 1900).

24. Henry Calderwood, "Ethical Aspects of the Theory of Development," *Contemporary Review* 31 (1877): 123–132, on pp. 124, 127. He further advanced this line of argument in his critique of Herbert Spencer. See Henry Calderwood, "Herbert Spencer on the Data of Ethics," *Contemporary Review* 37 (1880): 64–76; Henry Calderwood, "The Relations of Moral Philosophy to Speculation Concerning the Origin of Man," *Princeton Review* 8 (1881): 288–302.

25. Henry Calderwood, "Evolution, Physical and Dialectic," *Contemporary Review* 40 (1881): 865–876, on pp. 867–868.

26. Henry Calderwood, "On Evolution and Man's Place in Nature," *Proceedings of the Royal Society of Edinburgh* 17 (1889–1890): 71–79, on pp. 71, 74.

27. Henry Calderwood, *Evolution and Man's Place in Nature* (London: Macmillan, 1893), on pp. 2, 340.

28. Henry Calderwood, *Evolution and Man's Place in Nature*, 2nd ed. (London: Macmillan, 1896), p. 33.

29. Henry Calderwood, *The Relations of Science and Religion* (New York: Wilbur B. Ketcham, 1881), pp. 21, 134–135.

30. Calderwood, *Relations of Science and Religion*, pp. 153–154.

31. Calderwood, *Relations of Science and Religion*, pp. 237, 239, 277.

32. George D, Mathews, ed., *Alliance of the Reformed Churches Holding the Presbyterian System. Minutes and Proceedings of the Third General Council, Belfast 1884* (Belfast: Assembly's Offices, 1884), pp. 250–251.

33. Henry Calderwood, "Charles Darwin: His Boyhood," *United Presbyterian Magazine* (1888): 15–18; "Charles Darwin: His Youth," *United Presbyterian Magazine* (1888): 58–60; "Darwin Under Weigh as a Naturalist," *United Presbyterian Magazine* (1888): 113–116; and "Charles Darwin: His Theory as to Coral Reefs," *United Presbyterian Magazine* (1888): 254–257.

34. George Matheson, *Can the Old Faith Live with the New? Or, the Problem of Evolution and Revelation* (Edinburgh: William Blackwood & Sons, 1885). See also George Matheson, "The Religious Bearings of the Doctrine of Evolution," in *Alliance of The Reformed Churches Holding the Presbyterian System: Minutes and Proceedings of the Third General Council, Belfast 1884* (Belfast: Assembly's Offices, 1884). Matheson is briefly discussed in James R. Moore, *The Post-Darwinian Controversies: A Study of the Protestant Struggle to Come to Terms with Darwin in Great Britain and America, 1870–1900* (Cambridge: Cambridge University Press, 1979), pp. 228–229. See also D. Macmillan, *The Life of George Matheson* (London: Hodder and Stoughton, 1907); W. F. Gray, "Matheson, George (1842–1906)," rev. H. C. G. Matthew, *Oxford Dictionary of National Biography* (Oxford: Oxford University Press, 2004).

35. Matheson, *Can the Old Faith Live with the New?* pp. 3, 79, 191, 91, 168, 106, 377, 380, 387.

36. George Matheson, "Modern Science and the Religious Instinct," *Presbyterian Review* 5 (1884): 608–621, on pp. 608, 609, 613, 617, 618.

37. Iverach's stance is discussed in Moore, *Post-Darwinian Controversies*, pp. 253–259; and David N. Livingstone, *Darwin's Forgotten Defenders: The Encounter between Evangelical Theology and Evolutionary Thought* (Edinburgh: Scottish Academic Press, 1987), pp. 138–140. An overview of Iverach's life and work is available in Alan P. F. Sell, *Defending and Declaring the Faith: Some Scottish Examples, 1860–1920* (Exeter: Paternoster Press, 1987), chapter 5, "James Iverach (1839–1922): Theologian at the Frontier."

38. James Iverach, *Christianity and Evolution* (London: Hodder and Stoughton, 1894), pp. 18, 75, 113, 109, 110, 107.

39. Iverach, *Christianity and Evolution*, pp. 3, 79, 95, 101, 121.

40. The theological possibilities of the idea of evolutionary cooperation have recently resurfaced. For example, Sarah Coakley's 2012 Gifford lectures, entitled "Sacrifice Regained: Evolution, Cooperation and God," argued for a reconstituted natural theology and were based, in part, on her work with the Harvard mathematical biologist Martin Nowak. Nowak, a Catholic, has himself proposed a partnership between science and religion. See Rich Barlow, "Mathematics and Faith Explain Altruism," *Boston Globe*, 27 Sept. 2008. For evidence of a resurgence of interest in scientific dimensions of this theme, now couched in the language of mathematical modeling, see Martin A. Nowak, Robert M. May, and Karl Sigmund, "The Arithmetics of Mutual Help: Computer Experiments Show How Cooperation Rather than Exploitation Can Dominate in the Darwinian Struggle for Survival," *Scientific American* 272 (June 1995): 76–81; Karl Sigmund, Ernst Fehr, and Martin A. Nowak, "The Economics of Fair Play," *Scientific American* 286 (Jan. 2002): 82–87; and, more recently, Christoph Adami and Arend Hintze, "Evolutionary Instability of Zero-Determinant Strategies Demonstrates that Winning is Not Everything," *Nature Communications* 4 (2013), online at http://www.nature.com/ncomms/2013/130801/ncomms3193/pdf/ncomms3193.pdf.

41. Iverach, *Christianity and Evolution*, pp. 114, 181, 183, 127.

42. Iverach, *Christianity and Evolution*, pp. 207, 226.

43. James R. Moore, "Evangelicals and Evolution: Henry Drummond, Herbert Spencer, and the Naturalisation of the Spiritual World," *Scottish Journal of Theology* 38 (1985): 383–417, on p. 404.

44. Several biographical treatments are available, including Cuthbert Lennox, *Henry Drummond: A Biographical Sketch* (London: Andrew Melrose, 1901); James Young Simpson,

Henry Drummond (Edinburgh: Oliphant Anderson & Ferrier, 1901); George Adam Smith, *The Life of Henry Drummond* (London: Hodder and Stoughton, 1902).

45. This was taken up by Ernst Benz in *Evolution and Christian Hope: Man's Concept of the Future from the Early Fathers to Teilhard de Chardin*, trans. from the German by Heinz G. Frank (New York: Doubleday, 1966).

46. Simpson, *Henry Drummond*, p. 100.

47. Lennox, *Henry Drummond*, pp. 71–73. Anne Scott has analyzed the form and function of the original lectures as a contribution to popular apologetics in Anne Scott, " 'Visible Incarnations of the Unseen': Henry Drummond and the Practice of Typological Exegesis," *British Journal for the History of Science* 37 (2004): 435–454.

48. Moore, "Evangelicals and Evolution," p. 386.

49. Henry Drummond, *Natural Law in the Spiritual World* (London: Hodder and Stoughton, 1883), pp. 11, 18, 35.

50. It might be seen as a contribution to the "sermons from nature" tradition, which looked for the hand of divinity in the wonders of creation. On a comparable literary genre, see Aileen Fyfe, *Science as Salvation: Evangelical Popular Science Publishing in Victorian Britain* (Chicago: University of Chicago Press, 2004).

51. Drummond, *Natural Law*, pp. 75, 120. Scott examines the wider context of the use of typology and analogy in works of evangelical instruction at the time. Scott, " 'Visible Incarnations of the Unseen.' "

52. Simpson, *Henry Drummond*, p. 143.

53. Henry Drummond, *The Ascent of Man* (London: Hodder and Stoughton, 1894), p. vi.

54. See Peter J. Bowler, *The Eclipse of Darwinism: Anti-Darwinian Theories of Evolution in the Decades Around 1900* (Baltimore: Johns Hopkins University Press, 1983).

55. Drummond, *Ascent of Man*, pp. 8, 11, 13.

56. Drummond, *Ascent of Man*, pp. 275, 281, 24, 39, 435, 37.

57. Later, Kropotkin referred to Drummond's contributions. See Petr Kropotkin, *Mutual Aid: A Factor of Evolution* (Harmondsworth: Pelican, 1939), p. 19.

58. Drummond, *Ascent of Man*, p. 438.

59. Cited in Moore, "Evangelicals and Evolution," p. 399.

60. Moore, "Evangelicals and Evolution," p. 401.

61. Robert Watts, *Professor Drummond's "Ascent of Man," and Principal Fairbairn's "Place of Christ in Modern Theology" Examined in the Light of Science and Revelation* (Edinburgh: R. W. Hunter, [1894]), pp. vi, 81.

62. Cited in Moore, "Evangelicals and Evolution," p. 399.

63. Lennox, *Henry Drummond*, p. 174.

64. W. Robertson Nicoll's tribute to Drummond appeared in the *British Weekly*, 18 Mar. 1987, and is reprinted in *Princes of the Church* (London: Hodder and Stoughton, 1921), pp. 93–102, quotation on p. 96. His comment to Dods may be found in T. H. Darlow, *William Robertson Nicoll: Life and Letters* (London: Hodder and Stoughton, 1925), p. 163.

65. Aubrey L. Moore, "Prof. H. Drummond's 'Natural Law in the Spiritual World,' " in *Science and the Faith: Essays on Apologetic Subjects* (London: Kegan Paul, Trench, Trübner, 1893), pp. 1–29, on pp. 1, 2.

66. Quoted in Michael Jinkins, "Bruce, Alexander Balmain (1831–1899)," *Oxford Dictionary of National Biography* (Oxford: Oxford University Press, 2004).

67. Alexander Balmain Bruce, *The Providential Order of the World* (London: Hodder and Stoughton, 1897), pp. 14, 15, 26, 27, 48, 60–61.

68. Aubrey Moore's comments originally appeared in Moore, "Darwinism and the Christian Faith," in *Science and the Faith*, pp. 184, 185. Bruce's quotation of Moore appeared in *The Providential Order*, p. 60.

69. Orthogenesis was an evolutionary theory contending that organisms have an inherent tendency, driven by internal factors, to develop in predetermined directions. The mutation theory proposed that evolution took place by intermittent, sudden, large-scale variations that cast doubt on the gradualism of Darwinian mechanisms. Neo-Lamarckism returned to Lamarck's ideas about the inheritance of characteristics acquired in response to environmental stimuli but combined this commitment with an acknowledgement of the role played by natural selection. See Bowler, *Eclipse of Darwinism*.

70. I have discussed Orr's perspective in *Darwin's Forgotten Defenders*, pp. 140–144. See also Glen G. Scorgie, *A Call for Continuity: The Theological Contribution of James Orr* (Macon, GA: Mercer University Press, 1988); David W. Bebbington, "Orr, James (1844–1913)," *Oxford Dictionary of National Biography* (Oxford: Oxford University Press, 2004).

71. See James Orr, *The Ritschlian Theology and the Evangelical Faith* (London: Hodder and Stoughton, 1898).

72. James Orr, *The Christian View of God and the World* (Edinburgh: Andrew Elliott, 1893).

73. James Orr, *God's Image in Man and its Defacement* (London: Hodder and Stoughton, 1905), pp. 126, 152, 153.

74. James Orr, *The Faith of a Modern Christian* (London: Hodder and Stoughton, 1910), p. 215.

75. Orr, *God's Image in Man*, pp. 95–96.

76. James Orr, *The Progress of Dogma*, 4th ed. (1897; London: Hodder and Stoughton, 1901), pp. 17–18, 19.

77. The persistence of an anti-Darwinian outlook among those conservatives who later broke away to form the Free Presbyterian Church of Scotland in 1893—the so-called second Disruption—is recorded in James Lachlan MacLeod, *The Second Disruption: The Free Church in Victorian Scotland and the Origins of the Free Presbyterian Church* (East Linton: Tuckwell Press, 2000), pp. 110–123. Much of MacLeod's evidence for this opposition comes from reviews of works by figures like Drummond that appeared in the new denomination's magazine during the late 1890s.

78. Robert Rainy, *Faith and Science. A Sermon Preached in St. George's Free Church, Edinburgh . . . in Connection with the University of Edinburgh Tercentenary Celebration* (Edinburgh: Macniven and Wallace, 1884), p. 13.

79. James Iverach, *The Christian Message and Other Lectures* (London: Hodder and Stoughton, 1920), p. 56.

80. The most recent biography is Bernhard Maier, *William Robertson Smith: His Life, His Work, His Times* (Tübingen: Mohr Siebeck, 2009).

81. These, with a variety of other essays, are reprinted in John Sutherland Black and George Chrystal, eds., *Lectures and Essays of William Robertson Smith* (London: Adam and Charles Black, 1912).

82. William Robertson Nicoll, in an obituary notice for the *British Weekly*, observed that Smith's "appointment to a theological chair over the heads of ordained ministers [was] a thing previously unheard of." Nicoll, *Princes of the Church*, p. 65.

83. Key figures in this tradition on whom Smith relied were Julius Wellhausen, Wilhelm Vatke, Abraham Keunen, and Karl Heinrich Graf. Their contributions and others' are discussed in John W. Rogerson, *Old Testament Criticism in the Nineteenth Century* (London: S.P.C.K., 1984).

84. William Robertson Smith, "Bible," *Encyclopaedia Britannica*, 9th ed., vol. 3 (1875), p. 634, 636.

85. Archibald Hamilton Charteris, "The New *Encyclopaedia Britannica* on Theology," *Edinburgh Courant*, 16 Apr. 1877.

86. Rogerson has pointed to the political elements in the debacle. He argues that Smith's article disturbed the biblical ground underlying the political philosophy of the well-known Free Church preacher James Begg. That philosophy had encouraged Begg to devote his energies to working-class housing and Scottish nationalism. See John W. Rogerson, *The Bible and Criticism in Victorian Britain: Profiles of F. D. Maurice and William Robertson Smith* (Sheffield: Sheffield Academic Press, 1995), pp. 56–65.

87. Besides his entries for the *Encyclopaedia Britannica*, several of Smith's other writings were the subject of censure, including William Robertson Smith, "The Question of Prophecy in the Critical Schools of the Continent," *British Quarterly* (Apr. 1870), reprinted in Black and Chrystal, *Lectures and Essays*, pp. 163–203; and William Robertson Smith, "The Sixteenth Psalm," *Expositor* 23 (Nov. 1876): 341–372.

88. John Sutherland Black and George Chrystal, *The Life of William Robertson Smith* (London: Adam and Charles Black, 1912), pp. 406–407. On Smith's travels in the Middle East, see James A. Thrower, "Two Unlikely Travelling Companions: Sir Richard Burton and William Robertson Smith in Egypt 1880," in Johnstone, *William Robertson Smith*, pp. 383–398. I have discussed the significance of Smith's expeditions in David N. Livingstone, "Oriental Travel, Arabian Kinship and Ritual Sacrifice: William Robertson Smith and the Fundamental Institutions," *Society and Space* 22 (2004): 639–657.

89. William Robertson Smith, "Prophecy and Personality: A Fragment" (Jan. 1868), printed in Black and Chrystal, *Lectures and Essays*, pp. 97–108, on pp. 97, 98. The influence of Kantian idealism on Smith is noted in Joachim Schaper, "William Robertson Smith's Early Work on Prophecy and the Beginnings of Social Anthropology," *Journal of Scottish Thought* 1 (2008): 13–23.

90. J. F. McLennan, "The Worship of Animals and Plants," *Fortnightly Review*, n.s., 4 (1869): 407–427, 562–582; 7 (1870): 194–216. See the discussion in Adam Kuper, *The Invention of Primitive Society: Transformations of an Illusion* (London: Routledge, 1991), pp. 82–83; Robert Alun Jones, *The Secret of the Totem: Religion and Society from McLennan to Freud* (New York: Columbia Press, 2005).

91. John F. McLennan, *Primitive Marriage: An Inquiry into the Origins of the Form of Capture in Marriage Ceremonies* (Edinburgh: A&C Black, 1865).

92. On the importance of pre-Darwinian Victorian social evolutionism, see the still valuable account by John W. Burrow, *Evolution and Society: A Study in Victorian Social Theory* (Cambridge: Cambridge University Press, 1966). More generally, see George W. Stocking, *Victorian Anthropology* (New York: Free Press, 1987); Peter J. Bowler, *The Invention of Progress: The Victorians and the Past* (Oxford: Blackwell, 1989); Stephen K. Sanderson, *Social Evolutionism: A Critical History* (Oxford: Blackwell, 1990); A. Bowdoin Van Riper, *Men Among the Mammoths: Victorian Science and the Discovery Of Human Prehistory* (Chicago: Univer-

sity of Chicago Press, 1993); and William Y. Adams, *The Philosophical Roots of Anthropology* (Stanford: CSLI Publications, 1998).

93. On the subject of anthropological "survivals," see Margaret Hodgen, *The Doctrine of Survivals: A Chapter in the History of Scientific Method in the Study of Man* (London: Allenson, 1936).

94. Some of the connections between McLennan and Smith are discussed in Peter Rivière, "William Robertson Smith and John Ferguson McLennan: The Aberdeen Roots of British Social Anthropology" in William Johnstone, ed., *William Robertson Smith: Essays in Reassessment* (Sheffield: Sheffield Academic Press, 1995), pp. 293–302.

95. William Robertson Smith, "Animal Worship and Animal Tribes among the Arabs and in the Old Testament," *Journal of Philology* 9 (1880): 75–100, on p. 78. Smith further developed this whole line of thinking in William Robertson Smith, *Kinship and Marriage in Early Arabia* (Cambridge: Cambridge University Press, 1885).

96. *Scotsman*, 29 July 1880, p. 4.

97. Cited in Black and Chrystal, *The Life of William Robertson Smith*, pp. 381–382.

98. Black and Chrystal, *The Life of William Robertson Smith*, p. 437.

99. The symbolic significance of this episode has been noted on numerous occasions. See, for instance, Don Cupitt, *The Sea of Faith* (London: BBC, 1995), and Peter Hinchliff, *God and History: Aspects of British Theology, 1875–1914* (Oxford: Clarendon Press, 1992).

100. Charles E. Raven, "William Robertson Smith. Oration delivered in the Elphinstone Hall, King's College, University of Aberdeen, 8th November 1946," printed in *Centenary of the Birth on 8th November 1846 of the Reverend Professor W. Robertson Smith* (Aberdeen: The University Press, 1951), p. 6.

101. S. A. Cook, "William Robertson Smith. Oration delivered in Christ's College, Aberdeen," printed in *Centenary*, p. 11.

102. William Johnstone comments that "Smith maintained . . . to his dying day, that he was an evangelical Christian who had said nothing that contradicted the [Westminster] Confession [of Faith]. His life's work was to defend the reality of divine revelation in Scripture." See William Johnstone, introduction to Johnstone, *William Robertson Smith*, pp. 15–22, on p. 20.

103. Black and Chrystal, *The Life of William Robertson Smith*, p. 210.

104. W. E. Henley, "Modern Men—Mr Robertson Smith," *Scots Observer*, 25 May 1889.

105. Patrick Carnegie Simpson, *The Life of Principal Rainy* (London: Hodder and Stoughton, [circa 1909]), 1:396.

106. Cited in Rogerson, *The Bible and Criticism*, p. 60.

107. "The Ascent of Man," *Echo*, 19 June 1894. I am most grateful to Mark Wood for drawing this comment and the one following to my attention.

108. "The Lowell Lectures on the Ascent of Man," *Aberdeen Free Press*, 4 June 1894.

109. Cited in Black and Chrystal, *The Life of William Robertson Smith*, pp. 80–81, 333.

110. This whole episode is treated in detail in chapter 3.

111. Crosbie Smith, "P. G. Tait, Queen's College and Ulster-Scots Natural Philosophy," in Alvin Jackson and David N. Livingstone, eds., *Queen's Thinkers: Essays in the Intellectual Heritage of a University* (Belfast: Blackstaff Press, 2008), pp. 7–17, on p. 9.The dispute was over Tyndall's "alleged defamation of J.D. Forbes." See G. K. Booth, "William Robertson

Smith: The Scientific, Literary and Cultural Context from 1866 to 1881" (Ph.D. thesis, University of Aberdeen, 1999).

112. Quoted in Booth, "William Robertson Smith," chapter 5, "The "Rights of Matter." Available at www.gkbenterprises.fsnet.co.uk/thesis/cho5.htm

113. William Robertson Smith, Letter, *Northern Whig*, 22 Aug. 1874. All this was familiar territory to Smith, for he had collaborated with T. M. Lindsay (later principal of the Glasgow Free Church College) on a paper given to the Mathematics and Physics Section of the British Association on Lucretius and Democritus at its Edinburgh meeting in 1871.

114. Keith W. Whitelam, "William Robertson Smith and the So-Called New Histories of Palestine," in Johnstone, *William Robertson Smith*, pp. 180–189, on p. 181.

115. Smith, "Question of Prophecy," p. 165.

116. Smith, "Bible," pp. 634, 635.

117. William Robertson Smith, "Hebrew Language and Literature," *Encyclopaedia Britannica*, 9th ed., vol. 9, (1880), p. 596.

118. William Robertson Smith, "The Progress of Old Testament Studies," *British and Foreign Evangelical Review* 25 (1876): 471–493, on p. 489.

119. Gillian M. Bediako, *Primal Religion and the Bible: William Robertson Smith and His Heritage* (Sheffield: Sheffield Academic Press, 1997), p. 157.

120. Rogerson, *Old Testament Criticism*, p. 253. It is not insignificant, I think, that Thomas Henry Huxley used the New Testament equivalent—the Synoptic Problem and the resort to an ur-text—to support his evolutionary outlook. See "Agnosticism" and "Agnosticism: A Rejoinder" in Thomas Henry Huxley, *Science and Christian Tradition: Essays* (London: Macmillan, 1895). For a discussion of this whole subject, see Matthew Day, "Reading the Fossils of Faith: Thomas Henry Huxley and the Evolutionary Subtext of the Synoptic Problem," *Church History* 74 (2005): 534–556.

121. As Beidelman observes, "Smith was convinced of the evolutionary progress of both society and, more important, the intellectual consciousness that was both a cause and result of that changing social environment. Nearly all of his writings were cast in some developmental form, whether it was a report of the increasing refinement and progress of biblical scholarship or an account of the evolution of ancient Semitic kinship organisation." T. O. Beidelman, *W. Robertson Smith and the Sociological Study of Religion* (Chicago: University of Chicago Press, 1974), p. 29.

122. Christopher Herbert, *Culture and Anomie: Ethnographic Imagination in the Nineteenth Century* (Chicago: University of Chicago Press, 1991), p. 56.

123. William Robertson Smith, *Lectures on the Religion of the Semites. First Series. The Fundamental Institutions* (1889; London: Black, 1907), pp. 30, 34, 51, 52, 24.

124. Smith, *Lectures on the Religion of the Semites*, pp. 260, 16, 18.

125. Accounts of Smith's anthropology of sacrifice abound. See, e.g., Edmund Leach, "Anthropology of Religion: British and French Schools," in Ninian Smart, John Clayton, Patrick Sherry, and Steven T. Katz, eds., *Nineteenth Century Religious Thought in the West*, vol. 3 (Cambridge: Cambridge University Press, 1985), pp. 215–262; Margit Warburg, "William Robertson Smith and the Study of Religion," *Religion* 19 (1989): 41–61; Gordon Booth, "The Fruits of Sacrifice: Sigmund Freud and William Robertson Smith," *Expository Times* 113 (2002): 258–264; George Elder Davie, "Scottish Philosophy and Robertson Smith," in *The Scottish Enlightenment and Other Essays* (Edinburgh: Polygon, 1991), pp. 101–145. Numbered among leading anthropologists who have deployed versions of Smith's analysis are

Durkheim, Malinowski, Evans-Prichard, Hubert and Mauss, and Mary Douglas. See Emile Durkheim, *The Elementary Forms of Religion* (London: Allen and Unwin, 1915); Bronislaw Malinowski, *Magic, Science and Religion* (New York: Doubleday, 1954); E. E. Evans-Pritchard, *Neuer Religion* (Oxford: Clarendon Press, 1956); Henri Hubert and Marcel Mauss, *Sacrifice: Its Nature and Function* (London: Cohen & West, 1964), originally published as *Essai sur la Nature et la Fonction du Sacrifice* in *L'Année Sociologique* (1898), pp. 29–138; and Mary Douglas, *Purity and Danger* (London: Routledge, 1969).

126. Smith, *Lectures on the Religion of the Semites*, pp. 224, 269. See also William Robertson Smith, "Sacrifice," *Encyclopaedia Britannica*, 9th ed., vol. 21 (1886).

127. Smith, *Lectures on the Religion of the Semites*, p. 313.

128. William Robertson Smith, *Kinship and Marriage in Early Arabia* (1885; London: Adam and Charles Black, new ed., 1903), p. 306.

129. Davie, "Scottish Philosophy and Robertson Smith," pp. 101–145.

130. The influence of Smith's use of the idea of survivals or relics is noted in John W. Rogerson, *Anthropology and the Old Testament* (Oxford: Blackwell, 1978), pp. 25–28.

131. Smith, *Lectures on the Religion of the Semites*, pp. 345, 295, 428, 415.

132. Smith, *Lectures on the Religion of the Semites*, pp. 373, 439.

133. Sigmund Freud, *Totem and Taboo: Some Points of Agreement between the Mental Lives of Savages and Neurotics* (1913; London: Routledge, 2002), pp. 154, 163–164. Freud's biographer, E. Jones, noted that Freud "had hardly ever been so pleased with any book." Cited in Booth, "Fruits of Sacrifice," p. 259.

134. Freud, *Totem and Taboo*, p. 185.

135. Daniel Dennett, *Darwin's Dangerous Idea: Evolution and the Meanings of Life* (London: Penguin, 1996).

136. Letter to the *Scotsman*, 1 June 1877.

137. P. C. Simpson, *The Life of Principal Rainy* (London: Hodder and Stoughton, [n.d.]), p. 285.

Chapter 3 · Belfast, the Parliament of Science, and the Winter of Discontent

1. "The General Assembly's College, Belfast," *Belfast News-Letter*, 11 Nov. 1874. The full text soon appeared as a pamphlet: J. L. Porter, *Theological Colleges: Their Place and Influence in the Church and in the World; with Special Reference to the Evil Tendencies of Recent Scientific Theories. Being the Opening Lecture of Assembly's College, Belfast, Session 1874–75* (Belfast: William Mullan, 1874).

2. "The General Assembly's College, Belfast."

3. "Opening of the New College Edinburgh: Principal Rainy on the Evolution Theory," *Scotsman* 5 Nov. 1874, p. 3.

4. This was reported in the *Witness*, 28 Aug. 1874. Porter's scientific contributions are discussed in Edwin James Aiken, *Scriptural Geography: Portraying the Holy Land* (London: I. B. Tauris, 2009).

5. Henry Wallace, "Teachings of the British Association," manuscript held by the Gamble Library, Union Theological College, Belfast. The text of the lecture was published as "Teachings of the British Association," *Plain Words* 12 (Oct. 1874): 253–257.

6. Correspondence, *Belfast News-Letter*, 27 Aug. 1874.

7. Useful philosophical, political and theological background is provided in Andrew R. Holmes, "Presbyterians and Science in the North of Ireland before 1874," *British Journal*

for the History of Science 41 (2008): 541–565; Andrew R. Holmes, "Presbyterian Religion, Historiography, and Ulster Scots Identity, c. 1800 to 1914," *Historical Journal* 52 (2009): 615–640; and Andrew R. Holmes, "Covenanter Politics: Evangelicalism, Political Liberalism, and Ulster Presbyterians, 1798 to 1914," *English Historical Review* 125 (2010): 340–368.

8. See Andrew R. Holmes, "Professor James Thomson Senior and Lord Kelvin: Religion, Science, and Liberal Unionism in Ulster and Scotland," *Journal of British Studies* 50 (2011): 100–124.

9. J. David Hoeveler Jr., *James McCosh and the Scottish Intellectual Tradition: From Glasgow to Princeton* (Princeton: Princeton University Press, 1981). See also William Milligan Sloane, ed., *The Life of James McCosh: A Record Chiefly Autobiographical* (Edinburgh: T. & T. Clark, 1896).

10. James McCosh, "On the Method in Which Metaphysics Should be Prosecuted," printed in the *Belfast Mercury*, 13 Jan. 1852, and the *Belfast News Letter*, 14 Jan. 1852. I have discussed this in David N. Livingstone, "James McCosh and the Scottish Intellectual Tradition," in Alvin Jackson and David N. Livingstone, eds., *Queen's Thinkers: Essays on the Intellectual History of a University* (Belfast: Blackstaff Press, 2008), pp. 25–36.

11. James McCosh, *The Method of the Divine Government: Physical and Moral* (Edinburgh: Sutherland and Knox, 1850); James McCosh, "Morphological Analogy between the Disposition of the Branches of Exogenous Plants and the Venation of their Leaves," *Report of the 22nd Meeting of the British Association for the Advancement of Science, held at Belfast* (1852): 66–68; James McCosh, "Some Observations on the Morphology of Pines and Firs," *Report of the 24th Meeting of the British Association for the Advancement of Science, held at Liverpool* (1854): 99–100; James McCosh and George Dickie, *Typical Forms and Special Ends in Creation* (Edinburgh: Constable, 1856).

12. William Gibson, *Maynooth. A Protest Against its Endowment* (Belfast, 1843), p. 9. This is discussed in Holmes, "Presbyterians and Science."

13. James G. Murphy, *Science and Religion before the Flood* (Belfast: Shepherd and Aitchison, 1857), p. 29.

14. James Gerald Donat, "British Medicine and the Ulster Revival of 1859" (Ph.D. thesis, University of London, 1986), p. 13.

15. Rev. William MacIlwaine, *Revivalism Reviewed* (Belfast: T. M'Ilroy, 1859), pp. 15, 9. See also William MacIlwaine, "On Physical Affectations in Connection with Religion, as illustrated by 'Ulster Revivalism,'" *Journal of Mental Science* 6 (July 1860), 439–460.

16. Rev. James McCosh, *The Ulster Revival and Its Physiological Accidents. A Paper Read Before the Evangelical Alliance, September 22, 1859* (Belfast: C. Aitchison, 1859), p. 5.

17. G. Macloskie, "The Natural History of Man," *Ulster Magazine* 3 (1863): 217–237, on p. 230.

18. Macloskie's later writings, when he moved to America, are discussed in chapter 6. While still in Belfast, he wrote popular natural science pieces for the Presbyterian magazine *Plain Words* and published "The Silicified Wood of Lough Neagh, with Notes on the Structure of Coniferous Wood," *Proceedings of the Belfast Natural History and Philosophical Society* (14 Feb. 1872): 51–74.

19. Macloskie, "The Natural History of Man," p. 230.

20. William Todd Martin, *Our Church in its Relation to Progressive Thought* (Newry: James Burns, 1863), pp. 11, 10.

21. James R. Moore, "1859 and All That: Remaking the Story of Evolution-and-Religion,"

in Roger G. Chapman and Cleveland T. Duval, eds., *Charles Darwin, 1809–1882: A Centennial Commemorative* (Wellington, New Zealand: Nova Pacifica, 1982), pp. 167–194, on p. 190.

22. Watts's lecture is reported in "General Assembly's College. Installation of Professor Watts," *Banner of Ulster*, 15 Nov. 1866.

23. See Charles Frederick D'Arcy, *The Adventures of a Bishop, A Phase of Irish Life: A Personal and Historical Narrative* (London: Hodder & Stoughton, 1934), pp. 82–83. On the role of the Natural History and Philosophical Society at the time, see Ruth Bayles, "Understanding Local Science: The Belfast Natural History Society in the Mid-Nineteenth Century," in David Attis and Charles Mollan, eds., *Science and Irish Culture* (Dublin: Royal Dublin Society, 2004), 139–169.

24. *Northern Whig*, 19 Nov. 1866.

25. Charles Darwin, *The Variation of Animals and Plants under Domestication*, vol. 2, popular edition (London: John Murray, 1905), p. 256–257.

26. Sir Arthur Keith, "The Adaptational Machinery Concerned in the Evolution of Man's Body," 12th Huxley Memorial Lecture, *Nature*, supplement, no. 2807 (18 Aug. 1923): 257–268, on p. 259; and *Concerning Man's Origin. Being the Presidential Address Given at the Meeting of the British Association Held in Leeds on August 31, 1927, and Recent Essays on Darwinian Subjects* (London: Watts & Co., 1927), p. 27. See also Charles Frederick D'Arcy, *Adventures of a Bishop*, p. 82; and *Providence and the World-Order*, The Alexander Robertson Lectures Delivered before the University of Glasgow, 1932 (London: Hodder & Stoughton, 1932), pp. 73–74.

27. Joseph John Murray, "The Origin of Organs of Flight," *Proceedings of the Belfast Naturalists' Field Club*, 7th Report (1869–1870).

28. Joseph John Murphy, *Habit and Intelligence in their Connexion with the Laws of Matter and Force: A Series of Scientific Essays*, 2 vols. (London: Macmillan, 1869), 1:294, 298, 308, 348. Murphy reportedly considered the second edition of this work, which was a substantial expansion of the first, to be his most valuable contribution to knowledge. Richard W. Seaver, "Joseph John Murphy," in Arthur Deane, ed., *Belfast Literary Society, 1901–1901. Historical Sketch, With Memoirs of Some Distinguished Members* (Belfast: McCaw, Stevenson & Orr, 1902), pp. 109–110.

29. Joseph John Murphy, "Presidential Address on the Present State of the Darwinian Controversy," *Proceedings of the Belfast Natural History and Philosophical Society* (1873–1874): 1–24, on p. 1.

30. John Joseph Murphy, "Presidential Address on Some Questions in Cosmological Science," *Proceedings of the Belfast Natural History and Philosophical Society* (1872–1873): 1–19, on p. 19.

31. Murphy, "Present State of the Darwinian Controversy," pp. 1, 4, 24.

32. Here I draw on my "Darwin in Belfast: The Evolution Debate," in John W. Foster, ed., *Nature in Ireland: A Scientific and Cultural History* (Dublin: Lilliput Press, 1997), 387–408; and "Darwinism and Calvinism: The Belfast-Princeton Connection," *Isis* 83 (1992): 408–428.

33. *Northern Whig*, 19 Aug. 1874.

34. "Professor Tyndall's Address at the British Association," *Witness*, 21 Aug. 1874.

35. Adrian Desmond and James Moore, *Darwin* (London: Michael Joseph, 1991), p. 611.

36. "Professor John Tyndall, D.C.L., LL.D," *Northern Whig*, 19 Aug. 1874.

37. "Our Visitors: Professor Tyndall," *Witness*, 21 Aug. 1874.

38. John Tyndall, *Address Delivered Before the British Association Assembled at Belfast* (London: Longmans, Green and Co., 1874), pp. 59, 61.

39. Tyndall's "materialism" is treated in Ruth Barton, "John Tyndall, Pantheist: A Re-reading of the Belfast Address," *Osiris*, 2nd ser., 3 (1987): 111–134; S. Kim, *John Tyndall's Transcendental Materialism and the Conflict between Science and Religion in Victorian England* (Lewiston, NY: Mellen University Press, 1996); Bernard Lightman, "Scientists as Materialists in the Periodical Press: Tyndall's Belfast Address," in Geoffrey Cantor and Sally Shuttleworth, eds., *Science Serialized: Representation of the Sciences in Nineteenth Century Periodicals* (Boston: M.I.T. Press, 2004).

40. George Bernard Shaw, *Man and Superman* (Westminster: Archibald Constable, 1903), Act IV, p. 164.

41. Lightman, "Scientists as Materialists in the Periodical Press," pp. 199–237.

42. Robert Watts, "An Irenicum: Or, a Plea of Peace and Co-operation between Science and Theology," in *The Reign of Causality: A Vindication of the Scientific Principle of Telic Causal Efficiency* (Edinburgh: T. & T. Clark, 1888), p. 3.

43. *Witness*, 9 Oct. 1874.

44. *Northern Whig*, 25 Aug. 1874, p. 8. See also *Witness*, 9 Oct. 1874.

45. Robert Watts, "Atomism—An Examination of Professor Tyndall's Opening Address before the British Association, 1874," in *The Reign of Causality*, pp. 27, 28, 36, 37. The *Ulster Echo* similarly balked at the moral implications of the theory. "Truth *versus* Error," *Ulster Echo*, 27 Aug. 1874.

46. "Rev. Dr. Watts on Prof. Tyndall's Address," *Witness*, 28 Aug. 1874.

47. *Witness*, 18 Sept. 1874.

48. Robert Watts, review of *Atomism: Dr. Tyndall's Atomic Theory of the Universe Examined and Refuted*, in *Witness*, 11 Sept. 1874.

49. The public challenge appeared in *Northern Whig*, 27 Aug. 1874, p. 8. See also Watts's letter to the *Witness*, 28 Aug. 1874. Of course, not all Presbyterians supported this course of action. "An Orthodox Presbyterian" wrote to express "deep indignation at the ridiculous and undignified position in which [the Presbyterian] Church has been placed by the action of her self-constituted champion, Dr. Watts. . . . Could the force of absurdity go further? Is it possible that the monstrous egotism which inspired such a proposal can escape the perception even of the simplest-minded among the orthodox admirers of Dr. Watts?" *Northern Whig*, 28 Aug. 1874, p. 8.

50. R. Jeffrey, "Scientific Giants v. Theological Pigmies," *Christian Banner* 2, no. 5 (Oct. 1874): 43–45, on p. 43.

51. William Macloy, "Professor Tyndall's Theory of Life and Organization," *Christian Banner* 2, no. 5 (Oct. 1874): 41–43, on p. 42.

52. Tyndall's address was also the subject of Catholic sermonizing that Sunday. According to the *Belfast Morning Post* (25 Aug. 1874), "At St. Patrick's, the Rev. M. H. Cahill delivered an eloquent sermon, during the course of which he combated the theories of Professor Tyndall, and argued that they were incompatible not only with the laws of Christianity, but the laws of science. . . . In other Catholic churches similar references were made." I am grateful to Diarmid Finnegan for drawing this report to my attention. More generally on this subject, see Diarmid Finnegan, "Catholics, Science and Civic Culture in Belfast," unpublished manuscript.

53. *Northern Whig*, 24 Aug. 1874, p. 8.

54. Charles Parsons Reichel, *The Necessary Limits of Christian Evidences: A Sermon Preached at Carnmoney Church & St. Thomas' Church, Belfast, on Sunday, August 23, 1874, during the Meeting of the British Association, and with Special Reference to the Inaugural Address of its President* (Belfast: W. H. Greer, 1874); James C. Street, *Science and Religion: A Sermon Preached in the Church of the Second Congregation, Belfast, During the Meetings of the British Association, on Sunday, August 23, 1874* (London: E. T. Whitfield, 1874); J. Scott Porter, *"The Existence and Attributes of the Invisible God Made Manifest Through his Visible Works." A Discourse Delivered on the 30th August, 1874 in the First Presbyterian Church, Rosemary Street, Belfast; In Which some Statements made by Professor Tyndall in his Inaugural Address, Read at the Annual Meeting of the British Association for the Advancement of Science, are Briefly Considered* (Belfast: Wm Henry Greer, 1874).

55. Cited in Ruth Barton, "John Tyndall, Pantheist. A Rereading of the Belfast Address," *Osiris*, 2nd ser., 3 (1987): 116.

56. The entry in the Committee Minute Book of the Rosemary Street Presbyterian Church for 3 September 1874 notes, "the Committee cannot depart without expressing their thanks to the Rev John MacNaughtan and . . . recording their high appreciation of the promptness and ability with which he replied to the materialistic doctrines and erroneous conclusions which were gratuitously advanced and covertly advocated by Professor Tyndall."

57. J. MacNaughtan, *The Address of Professor Tyndall, at the Opening of the British Association for the Advancement of Science, Examined in A Sermon on Christianity and Science* (Belfast: Aitchison, Reed, & Henderson, 1874). All this, of course, was for popular consumption, as was recognized at the time; according to one observer, the tract lacked "that thoroughness and closeness of treatment such an address as that of Professor Tyndall's requires for its final and complete refutation." See Robert Carswell, review of MacNaughtan's *Christianity and Science* in the *Witness*, 18 Sept. 1874.

58. On the role of rhetoric more generally in science, see Alan G. Gross, *The Rhetoric of Science* (Cambridge: Harvard University Press 1996); and for a particularly telling case study, see J. Vernon Jensen, "Return to the Wilberforce-Huxley Debate," *British Journal for the History of Science* 21 (1988): 161–179.

59. *McComb's Presbyterian Almanack, and Christian Remembrancer for 1875* (Belfast: James Cleeland, 1875), p. 84.

60. William MacIlwaine, "Presidential Address," *Proceedings of the Belfast Naturalists' Field Club* (Winter 1874–1875), 81–99, on pp. 82, 83, 84.

61. William Todd Martin, "The Recent Meeting of the British Association," manuscript held in the Gamble Library of the Union Theological College, Belfast.

62. "The British Association," *Witness*, 28 Aug. 1874. It should not be assumed that there were no less strident voices. Rev. George Macloskie, in a letter to the *Northern Whig* on 26 August, for example, insisted that there was no desire in Belfast "to stifle free scientific inquiry." *Northern Whig*, 27 Aug. 1874, p. 8.

63. This was also noted by the Dean of Manchester, who was "sorry to say" that Tyndall's Belfast address had been given "without any protest on the part of the audience" (*Witness*, 4 Sept. 1874). The *Witness* was a weekly newspaper established in January 1874 by the Presbyterian Church in Ireland. Some general reflections on the intellectual scene in Belfast at the time are to be found in Peter Brooke, "Religion and Secular Thought,

1800–75," in J. C. Beckett et al., *Belfast: The Making of a City, 1800–1914* (Belfast: Appletree Press, 1983), pp. 111–128.

64. "The Belfast Presbytery and the British Association," *Witness*, 4 Sept. 1874.

65. "Winter Lectures in Belfast," *Witness*, 25 Sept. 1874.

66. *Science and Revelation: A Series of Lectures in Reply to the Theories of Tyndall, Huxley, Darwin, Spencer, Etc.* (Belfast: William Mullan; New York: Scribner, Welford and Armstrong, 1875).

67. See Nuala C. Johnson, "Grand Design(er)s: David Moore, Natural Theology and the Royal Botanic Gardens in Glasnevin, Dublin, 1838–1879," *Cultural Geographies* 14 (2007): 29–55; Nuala C. Johnson, *Nature Displaced, Nature Displayed: Order and Beauty in Botanical Gardens*, (London: I.B. Tauris, 2011).

68. J. L. Porter, *Science and Revelation: Their Distinctive Provinces. With a Review of the Theories of Tyndall, Huxley, Darwin, and Herbert Spencer* (Belfast: William Mullan, 1874), pp. 3–4, 5, 20, 22, 21. The essay was reprinted as the first chapter of *Science and Revelation*.

69. J. L. Porter's career as president of Queen's College is discussed in T. W. Moody and J. C. Beckett, *Queen's, Belfast, 1845–1949: The History of a University* (London: Faber & Faber, 1959), 1:290–320.

70. Porter, *Theological Colleges*, p.8.

71. William Todd Martin, *The Doctrine of An Impersonal God in its Effects on Morality and Religion* (Belfast: William Mullan, 1875), pp. 7, 17. Reprinted in Porter, *Science and Revelation*.

72. [Henry] Wallace, *Prayer in Relation to Natural Law* (Belfast: William Mullan, 1875).

73. The letter is attributed to Henry Thompson, an eminent London physician. Thompson had proposed that over a period of time one hospital ward should be the object of sustained prayer and the mortality rates subsequently compared with the past records and with those of other leading hospitals. See "The 'Prayer for the Sick': Hints Towards a Serious Attempt to Estimate Its Value," *Contemporary Review* 20 (July 1872): 205–210. See the discussion in Frank Miller Turner, "Rainfall, Plagues and the Prince of Wales: A Chapter in the Conflict of Religion and Science," *Journal of British Studies* 13 (1974): 46–65. Tyndall himself insisted that his concern was not with prayer in general but only with its "physical value." When prayer was supposed to produce "the precise effects caused by physical energy in the ordinary course of things," Tyndall reasoned, it was a legitimate subject for empirical analysis. John Tyndall, "On Prayer," *Contemporary Review* 22 (Oct. 1972): 764.

74. See Theodore M. Porter, *The Rise of Statistical Thinking, 1820–1900* (Princeton: Princeton University Press, 1986), p. 137; Francis Galton, "Statistical Inquiries into the Efficacy of Prayer," *Fortnightly Review* 18 (1872): 125–135. Galton originally offered this piece to the *Contemporary Review*, but it was rejected because the editor felt it might overly try the patience of the serial's clerical readership. See Karl Pearson, *The Life, Letters and Labours of Francis Galton* (Cambridge: Cambridge University Press, 1924), 2:131. This whole episode is discussed in Robert Bruce Mullin, "Science, Miracles, and the Prayer-Gauge Debate," in David C. Lindberg and Ronald L. Numbers, eds., *When Science and Christianity Meet* (Chicago: University of Chicago Press, 2003), pp. 203–224.

75. Wallace earlier attacked Tyndall's speech in "Teachings of the British Association," *Plain Words* (Oct. 1874): 253–257.

76. Desmond and Moore, *Darwin*, p. 613.

77. J. G. C., "Darwinism," *Irish Ecclesiastical Record* 9 (May 1873): pp. 337–361, on p. 346.

78. "Pastoral Address of the Archbishops and Bishops of Ireland," *Irish Ecclesiastical Record* 11 (Nov. 1874), pp. 49, 54.

79. *Atomism: Dr. Tyndall's atomic theory of the universe examined and refuted. To which are added, Humanitarianism accepts, provisionally, Tyndall's impersonal atomic deity; and a letter to the presbytery of Belfast; containing a note from the Rev. Dr. Hodge, and a critique on Tyndall's recent Manchester recantation, together with strictures on the late manifesto of the Roman Catholic hierarchy of Ireland in reference to the sphere of science* (Belfast 1875), pp. 34, 38, 39.

80. This is even clear from the title of his address, *Science and revelation: Their distinctive provinces.*

81. J. MacNaughtan, *The Address of Professor Tyndall, at the Opening of the British Association for the Advancement of Science, Examined in a Sermon on Christianity and Science* (Belfast 1874), p. 5.

82. See also Editorial, *Witness*, 6 Nov. 1874, p. 6.

83. "The Evangelical Alliance," *Irish Ecclesiastical Record* 11 (Jan. 1874): 224.

84. "Pastoral Address," pp. 64–65. The bishops and archbishops were referring to Joseph Priestley and commented that the Anglican archbishop of Dublin, Dr. Whately, advocated similar views.

85. George Matheson, "The Religious Bearings of the Doctrine of Evolution," in George D. Mathews, *Alliance of the Reformed Churches Holding the Presbyterian System. Minutes and Proceedings of the Third General Council, Belfast, 1884* (Belfast: Assembly's Offices, 1884), pp. 82, 88.

86. Jean Monod, "Evolutionism and the Facts of Nature and of Revelation," in Mathews, *Alliance of the Reformed Churches*, p. 90; Calderwood, "Discussion," in Mathews, *Alliance of the Reformed Churches*, p. 250. Other delegates were less sure. Certainly, there were those like Thomas Smith of Edinburgh who thought that while evolution was compatible with theism, it was "irreconcilable with the Scriptural statements with regard to the manner of Creation." Smith, "Discussion," in Mathews, *Alliance of the Reformed Churches*, p. 249.

87. Robert Watts, "Discussion," in Mathews, *Alliance of the Reformed Churches*, pp. 251, 252.

88. Watts observed: "My lecture in Perth on Herbert Spencer's Biological Hypothesis came off very well." Robert Watts [Dundee] to his wife, 26 Sept. 1877, in "Family Letters of Revd. Robert Watts, D.D., LL.D," compiled by his wife (typescript). I am grateful to Dr. R. E. L. Rodgers for making this typescript available to me. [Robert] Watts, *An Examination of Herbert Spencer's Biological Hypothesis* (Belfast: William Mullan, 1875), p. 7.

89. This argument is remarkably akin to the more formal version recently advanced by Alvin Plantinga in, inter alia, James K. Beilby, ed., *Naturalism Defeated? Essays on Alvin Plantinga's Evolutionary Argument Against Naturalism* (Ithaca: Cornell University Press, 2002).

90. W. Todd Martin, *The Evolution Hypothesis: A Criticism of the New Cosmic Philosophy* (Edinburgh: James Gemmell, 1887), pp. 9, 34–35, 8.

91. Robert Watts, "The Huxleyan Kosmogony," in *The Reign of Causality*, pp. 215–250, on p. 236.

92. Robert Watts, "Evolution and Natural History," in *The Reign of Causality*, pp. 285–317, on p. 287.

93. Henry Drummond, *Natural Law in the Spiritual World* (London: Hodder and Stoughton, 1883); idem, *The Lowell Lectures on the Ascent of Man* (London: Hodder and Stoughton, 1894).

94. The most penetrating analysis is provided in James R. Moore, "Evangelicals and Evolution: Henry Drummond, Herbert Spencer, and the Naturalization of the Spiritual World," *Scottish Journal of Theology* 38 (1985): 383–417. See also George Adam Smith, *The Life of Henry Drummond* (London: Hodder and Stoughton, 1902); Cuthbert Lennox, *Henry Drummond. A Biographical Sketch (With Bibliography)* (London: Andrew Melrose, 1901); James Young Simpson, *Henry Drummond* (Edinburgh: Oliphant Anderson & Ferrier, n.d., circa 1901).

95. Robert Watts, "Natural Law in the Spiritual World," *British and Foreign Evangelical Review* (1885) reprinted in *The Reign of Causality*, pp. 320, 329, 324, 328, 353–354.

96. Robert Watts, *Professor Drummond's "Ascent of Man," and Principal Fairbairn's "Place of Christ in Modern Theology,"* Examined in the Light of Science and Revelation (Edinburgh: R. W. Hunter, n.d., circa 1894), pp. 12, 13, 19.

97. See Andrew Holmes, "Biblical Authority and the Impact of Higher Criticism in Irish Presbyterianism, ca. 1850–1930," *Church History* 75 (2002): 343–373; Andrew Holmes, "The Common Sense Bible: Irish Presbyterianism, Samuel Davidson, and Biblical Criticism, c. 1800 to 1865," in Scott Mandelbrote and Michael Ledger-Lomas, eds., *Dissent and the Bible in Britain, c. 1650–1950* (New York: Oxford University Press, 2013), 176–204.

98. James G. Murphy, *A Critical and Exegetical Commentary on the Book of Genesis* (Edinburgh: T. & T. Clark, 1863), p. vii.

99. Robert Watts, *The Newer Criticism and the Analogy of the Faith. A Reply to Lectures by W. Robertson Smith, M.A., on the Old Testament in the Jewish Church* (Edinburgh: T. & T. Clark, 1882), pp. ix, x, xvi.

100. Robert Watts to B. B. Warfield, 5 Oct. 1893, Warfield Papers, Archives, Speer Library, Princeton Theological Seminary.

101. Robert Watts to B. B. Warfield, 13 Oct. 1890, Warfield Papers, Archives, Speer Library, Princeton Theological Seminary.

102. See Watts, *Newer Criticism*; Robert Watts, *The New Apologetic and Its Claims to Scriptural Authority* (Edinburgh: T. & T. Clark, 1890).

103. Cited in Robert Allen, *The Presbyterian College Belfast, 1853–1953* (Belfast: William Mullan, 1954), p. 182.

104. W. H. Green, review of *The Newer Criticism and the Analogy of the Faith* by Robert Watts, *Presbyterian Review* 3 (1882): 411–412.

105. Robert Watts to B. B. Warfield, 20 Feb. 1894, Warfield Papers, Archives, Speer Library, Princeton Theological Seminary.

106. A little biographical information is available in Brian Walker, *Sentry Hill: An Ulster Farm and Family* (Dundonald: Blackstaff Press, 1981); I. R. Crozier, *William Fee McKinney of Sentry Hill: His Family and Friends* (Coleraine: Impact Printing, 1985).

107. S. B. G. McKinney, *The Origin and Nature of Man*, rev. ed. (London: Hutchinson, 1898), p. 181. This work appeared in several editions. Other books by McKinney include *The Abolition of Suffering* (London, 1890); *Disease and Sin: A New Text-Book for Medical and Divinity Students* (London: Wyman & Sons, 1886); *The Revelation of the Trinity* (Edinburgh and London: Oliphant, Anderson & Ferrier, 1891).

108. McKinney, *Origin and Nature of Man*, pp. 182–183.

109. McKinney, *The Science and Art of Religion* (London: Kegan Paul, Trench & Co. 1888), pp. 35–36.

110. McKinney, *Origin and Nature of Man*, pp. 209, 211, 213.

111. J. R. Leebody, "The Theory of Evolution, and Its Relations," *British and Foreign Evangelical Review* 21 (1872): 1–35, on pp. 3, 7–8, 34–35.

112. J. R Leebody, "The Scientific Doctrine of Continuity," *British and Foreign Evangelical Review* 25 (1876): 742–774, on p. 769.

113. Leebody's later support for evolution is clear in "Evolution," *Witness*, 10 Oct. 1890, p. 5; "Results and Influence of Nineteenth-Century Science," *Witness*, 3 Nov. 1899, p. 3; "The Influence of Nineteenth-Century Science on Religious Thought," *Witness*, 6 Apr. 1900, p. 5.

114. J. R. Leebody, *Religious Teaching and Modern Thought: Two Lectures* (London: Henry Frowde, 1889), p. 19. He went on to argue that "the very success of Professor Huxley's efforts to account for the origin of various false religions on evolution principles brings into greater prominence the utter failure of his attempts to account in like manner for the origin of Christianity."

115. Leebody, "Theory of Evolution," p. 2.

116. Leebody, "Scientific Doctrine of Continuity," p. 773.

117. An Irish Graduate, "The Irish University Question," *Fraser's Magazine* (1872): 55–64, on pp. 60, 63. The author was Leebody who also published a history of his own college. J. R. Leebody, *A Short History of McCrea Magee College, Derry, During its First Fifty Years* (Londonderry: Printed at "Derry Standard" Office, 1915).

118. For background see T. W. Moody, "The Irish University Question of the Nineteenth Century," *History* 43 (1958): 90–109.

119. See Greta Jones, "Darwinism in Ireland," in Attis and Mollan, *Science and Irish Culture*, pp. 115–137.

120. Alexander Macalister, "Reviews and Bibliographical Notices. Works on Life and Organisation," *Dublin Quarterly Journal of Medical Science* 50 (1870): 112–132, on pp. 131–132, 129.

121. Alexander Macalister, *Man, Physiologically Considered*, Present-Day Tracts No. 38 (London: Religious Tract Society, 1886), p. 47.

122. Alexander Macalister, "The Body—The Temple of God," *Plain Words* 9 (1 May 1871): 137–140.

123. Alexander Macalister, "Reviews and Bibliographical Notices: The Descent of Man, and Selection in Relation to Sex," *Dublin Quarterly Journal of Medical Science* 52 (1871): 133–152, on pp. 142, 141, 150.

124. G. E. S[mith], "Alexander Macalister, 1844–1919," *Proceedings of the Royal Society of London*, series B, 94 (1922–23): xxxii–xxxix, on pp. xxxiv, xxxv.

125. See the discussion in Peter J. Bowler, *Theories of Human Evolution: A Century of Debate 1844–1944* (Oxford: Blackwell, 1986).

126. Greta Jones, "Contested Territories: Alfred Cort Haddon, Progressive Evolutionism and Ireland," *History of European Ideas* 24 (1998): 195–211.

127. Quoted in Smith, "Alexander Macalister," p. xxxviii.

128. Greta Jones, "Scientists Against Home Rule," in D. George Boyce and Alan O'Day, eds., *Defenders of the Union: A Survey of British and Irish Unionism Since 1801* (London: New York, 2001), pp. 188–208, on p. 192.

129. Quoted in Greta Jones, "Catholicism, Nationalism and Science," *Irish Review* 20 (1997): 47–61, on p. 50.

130. Macalister himself expressed his delight that in the modern university the "machinery for research is no longer absorbed in the maintenance of a semi-monastic set of men of a certain degree of learning" and reported on "the application of the methods and principles of evolution in the historical study of all phenomena" as one of the "great characteristics of modern critical philosophy" to which the university had contributed. Prof. [Alexander] Macalister, *Presbyterian Conference held Under the Auspices of the Presbytery of Belfast, October 1894* (Belfast: Printed at "The Witness," 1894), pp. 34–40.

131. Alexander Macalister, *Evolution in Church History* (Dublin: Hodges, Figgis, 1882), pp. 35, 10. Incidentally his son, Robert Alexander Stewart Macalister, the highly distinguished archaeologist who worked on both Palestine and Celtic subjects, applied the principle of evolution to the history of ecclesiastical vestments as well as authoring significant works on prehistory. See R. A. S. Macalister, *Ecclesiastical Vestments: Their Development and History* (London: Elliot Stock, 1896).

132. Alexander Macalister, "Henry Drummond," *Bookman* 12 (Apr. 1897). Quoted in Moore, "Evangelicals and Evolution," p. 390.

133. Alexander Macalister, review of *Theism in the Light of Present Science and Philosophy*, by James Iverach, *Critical Review* 10 (1900): 229–236, on p 233.

Chapter 4 · Toronto, Knox, and Bacon's Bequest

1. R. J. Helmstadter, "Wild, Joseph," *Dictionary of Canadian Biography*, vol. 13, available online at: http://www.biographi.ca/009004-119.01-e.php?&id_nbr=7145&&PHPSESSID=ks18vuen408uk3umrrstghfss1

2. "The Bond Street Pulpit. Dr Wild Ridicules the Modern Theory of Evolution," *Toronto World*, 9 June 1884.

3. "Dr Wild and Evolution," *Toronto World*, 13 June 1884.

4. "Dominion Methodist Church. Opening of the Basement. 'The Language of Stones,'" *Ottawa Free Press*, 27 Dec. 1875. See Robert Bruce Mullin, "Science, Miracles, and the Prayer-Gauge Debate," in David C. Lindberg and Ronald L. Numbers, eds., *When Science and Christianity Meet* (Chicago: University of Chicago Press, 2003), pp. 202–224.

5. "An Old Heathen Corpse. Dr Talmage's Laughable Description of Evolution," *Toronto World*, 25 Nov. 1884.

6. Citations appear in H. H. Langton, *Sir Daniel Wilson. A Memoir* (Toronto: Thomas Nelson & Sons, 1929), pp. 196, 197.

7. Quoted in Martin L. Friedland, *The University of Toronto: A History* (Toronto: University of Toronto Press 2002), p. 105.

8. Michael Gauvreau, *The Evangelical Century: College and Creed in English Canada from the Great Revival to the Great Depression* (Montreal: McGill–Queen's University Press, 1991), p. 128.

9. Carl Berger, *Science, God, and Nature in Victorian Canada* (Toronto: University of Toronto Press, 1983), p. 69; Suzanne Zeller, "Environment, Culture, and the Reception of Darwin in Canada, 1859–1909," in Ronald L. Numbers and John Stenhouse, eds., *Disseminating Darwinism: The Role of Place, Race, Religion, and Gender* (Cambridge: Cambridge University Press, 1999), pp. 91–122, on p. 92.

10. Something of the response of other religious traditions in English Canada more

generally may be gleaned from A. B. McKillop, *A Disciplined Intelligence: Critical Inquiry and Canadian Thought in the Victorian Era* (Montreal: McGill–Queen's University Press, 1979); Ruth Weir, "Darwin and the Universities in Canada," *Interchange* 14/4–15/1 (1983–84): 70–79.

11. Quoted in Susan Sheets-Pyenson, *John William Dawson: Faith, Hope, and Science* (Montreal: McGill–Queen's University Press, 1996), p. 130.

12. John William Dawson, "Review of Darwin *On the Origin of Species by Means of Natural Selection*," *Canadian Naturalist and Geologist* 5 (1860): 100–120, on p. 101.

13. Dawson, "Review," p. 112, 102, 103, 110, 114, 118, 120, 100.

14. John William Dawson, "The Present Aspect of Inquiries as to the Introduction of Genera and Species in Geological Time," *Canadian Monthly* 2 (1872): 154–156, on p. 154.

15. Quoted in Sheets-Pyenson, *John William Dawson*, p. 133.

16. John William Dawson, *The Story of the Earth and Man*, 7th ed. (London: Hodder and Stoughton, 1882), p. 339.

17. Dawson, *Story of the Earth and Man*, p. 396.

18. Dawson, *Story of the Earth and Man*, p. 334.

19. John William Dawson, *Some Salient Points in the Science of the Earth* (London: Hodder and Stoughton, 1893), pp. 188, 189.

20. Berger, *Science, God, and Nature*, p. 31.

21. Gauvreau, *Evangelical Century*, p. 128.

22. See Mélanie Desmeules, "Les Années Chicoutimiennes du Naturaliste Canadien," *Saguenayensia* 44, no. 3, (2002): 19–21.

23. Quoted in Luc Chartrand, Raymond Duchesne, and Yves Gingras, *Histoire des Sciences au Québec* (Montreal: Boréal, 1987), p. 178.

24. Berger, *Science, God, and Nature*, p. 21.

25. See Chartrand, *Histoire des Sciences au Québec*.

26. Zeller, "Environment," p. 94.

27. Quoted in Sheets-Pyenson, *John William Dawson*, p. 89.

28. John William Dawson, *Archaia; or, Studies of the Cosmology and Natural History of the Hebrew Scriptures* (Montreal: Dawson, 1860); Hugh Miller, *The Footprints of the Creator, or, The Asterolepis of Stromness* (London: Johnstone and Hunter, 1849); *The Testimony of the Rocks: Geology in Its Bearings on the Two Theologies, Natural and Revealed* (Edinburgh: T. Constable, 1857).

29. John William Dawson, *Eden Lost and Won: Studies of the Early History and Final Destiny of Man as Taught in Nature and Revelation* (London: Hodder and Stoughton, 1895), p. vi.

30. John William Dawson, *Modern Science in Bible Lands* (London: Hodder and Stoughton, 1888), p. 13.

31. Dawson, *Eden Lost and Won*, pp. 52, 101, 197.

32. Dawson, *Modern Science in Bible Lands*, p. 179.

33. John William Dawson, "Evolution in Education," *Princeton Review* 58 (1882): 233–248, on pp. 233, 234.

34. Dawson, "Evolution in Education," p. 243.

35. Dawson, "Evolution in Education," p. 244.

36. Dawson, *Modern Science in Bible Lands*, p. 24.

37. Dawson, *Story of the Earth and Man*, p. 318.

38. See John F. Connell, "From Creation to Evolution: Sir William Dawson and the Idea of Design in the Nineteenth Century," *Journal of the History of Biology* 16 (1983): 137–170; David N. Livingstone, *Darwin's Forgotten Defenders: The Encounter between Evangelical Theology and Evolutionary Thought* (Edinburgh: Scottish Academic Press, 1987).

39. See Peter J. Bowler, *The Eclipse of Darwinism: Anti-Darwinian Evolution Theories in the Decades Around 1900* (Baltimore: John Hopkins University Press, 1983).

40. John William Dawson, *Modern Ideas of Evolution as Related to Revelation and Science* (London: Religious Tract Society, 1890), pp. 227–228.

41. Daniel Wilson, "The President's Address," *Canadian Journal of Industry, Science, and Art*, n.s., 5 (Mar. 1860), 109–126, on p. 113.

42. Suzanne Zeller, *Inventing Canada: Early Victorian Science and the Idea of a Transcontinental Nation* (Toronto: University of Toronto Press, 1987), p. 272.

43. "1853 Memorial of the Natural History Society of Montreal to Lord Elgin," quoted in Zeller, *Inventing Canada*, p. 158.

44. Daniel Wilson, "The President's Address," *Canadian Journal of Industry, Science, and Art*, n.s., 6 (Mar. 1861): 101–120, on p. 102, 106.

45. Wilson, "President's Address" (1861), pp. 118, 119, 114.

46. Wilson, "President's Address" (1861), p. 115.

47. See the discussion in David N. Livingstone, *Adam's Ancestors: Race, Religion and the Politics of Human Origins* (Baltimore: Johns Hopkins University Press, 2008); William Stanton, *The Leopard's Spots: Scientific Attitudes toward Race in America, 1815–59* (Chicago: University of Chicago Press, 1960); Stephen Jay Gould, *The Mismeasure of Man* (Harmondsworth: Penguin, 1984); George W. Stocking Jr., *Victorian Anthropology* (New York: Free Press, 1987).

48. Adrian Desmond and James Moore, *Darwin's Sacred Cause: Race, Slavery and the Quest for Human Origins* (London: Allen Lane, 2009).

49. Wilson, "President's Address" (1861), pp. 118, 119.

50. Marinell Ash et al., *Thinking with Both Hands: Sir Daniel Wilson in the Old World and the New* (Toronto: University of Toronto Press, 1999); Carl Berger, "Wilson, Sir Daniel," *Dictionary of Canadian Biography* (Toronto: University of Toronto Press, 2000), q.v.

51. See W. Douglas Simpson, "Sir Daniel Wilson and the *Prehistoric Annals of Scotland*: A Centenary Study," *Proceedings of the Society of the Antiquaries of Scotland* 97 (1963–64): 1–9; Marinell Ash, " 'A Fine, Genial, Hearty Band': David Laing, Daniel Wilson and Scottish Archaeology," in A. S. Bell, ed., *The Scottish Antiquarian Tradition: Essays to Mark the Bicentenary of the Society of Antiquaries of Scotland and Its Museum, 1780–1980* (Edinburgh: John Donald, 1981), pp. 86–113.

52. Bruce G. Trigger, "*Prehistoric Man* and Daniel Wilson's Later Canadian Ethnology," in Ash et al., *Thinking with Both Hands*, pp. 81–100, on p. 83. See also Bruce G. Trigger, "Daniel Wilson and the Scottish Enlightenment," *Proceedings of the Society of Antiquaries of Scotland* 122 (1992): 55–75.

53. Daniel Wilson, *Prehistoric Man: Researches into the Origin of Civilisation in the Old and the New World*, 2 vols. (Cambridge: Macmillan, 1862), 2:357.

54. D[aniel] W[ilson], "The Unity of the Human Race," *Canadian Journal of Industry, Science, and Art*, n.s., 3 (1855): 229–231. See the discussion in Suzanne Zeller, " 'Merchants of Light': The Culture of Science in Daniel Wilson's Ontario, 1853–1892," in Ash et al., *Thinking with Both Hands*, pp. 115–138.

55. Wilson, *Prehistoric Man*, 1:46, 15, 139, ix.

56. See Bennett McCardle, " 'Heart of Heart': Daniel Wilson's Human Biology," in Ash et al., *Thinking with Both Hands*, pp. 101–114.

57. Daniel Wilson, "Relative Racial Brain-Weight and Size," in *The Lost Atlantis and Other Ethnographic Studies* (Edinburgh: David Douglas, 1892), pp. 339–402, on p. 340. See also chapter 7, "Hybridity and Heredity," pp. 307–338.

58. Wilson, *Prehistoric Man*, 2:458, 459.

59. See the discussions in Robert J. Richards, *Darwin and the Emergence of Evolutionary Theories of Mind and Behaviour* (Chicago: University of Chicago Press, 1987); Peter J. Bowler, *Theories of Human Evolution* (Oxford: Blackwell, 1987); Malcolm Jay Kottler, "Alfred Russel Wallace, the Origin of Man, and Spiritualism," *Isis* 65 (1974): 145–192.

60. Wilson, "President's Address," (1861), p. 116.

61. Letter reproduced in Langton, *Sir Daniel Wilson*, p. 89.

62. Zeller, " 'Merchant of Light,' " p. 125.

63. E[dward] J[ohn] C[hapman], review of *Archaia: Studies of the Cosmogony and Natural History of the Hebrew Scriptures, Canadian Journal of Industry, Science, and Art*, n.s., 5 (1860): 59–62, on p. 61.

64. For example, Edward J. Chapman, *Practical Mineralogy, or A Compendium of the Distinguishing Characteristics of Minerals, by Which the Name of any Species or Variety in the Mineral Kingdom May Be Speedily Ascertained* (London: H. Baillière, 1843); Edward J. Chapman, *A Brief Description of the Characters of Minerals: Forming a Familiar Introduction to the Science of Mineralogy* (London: H. Baillière, 1844); Edward J. Chapman, *A Popular and Practical Exposition of the Minerals and Geology of Canada* (Toronto: W. C. Chewett, 1864); Edward J. Chapman, *The Minerals and Geology of Central Canada: Comprising the Provinces of Ontario and Quebec with Explanatory and Technical Observations on Minerals, Rocks and Fossils Generally. A Handbook for Practical Use* (Toronto: Copp, Clark, 1871); Edward J. Chapman, *The Mineral Indicator: A Practical Guide to the Determination of Generally-occurring Minerals* (Toronto: Copp, Clark, 1884); Edward J. Chapman, *A Song of Charity* (Toronto: A. H. Armour, 1857).

65. E. J. C[hapman], review of *On the Origin of Species by Means of Natural Selection, Canadian Journal of Industry, Science, and Art*, n.s., 5 (1860): 367–387, on pp. 369, 387, 368.

66. C[hapman], review of *On the Origin of Species*, p. 369, 374, 379, 376, 387, 386.

67. E. J. C[hapman], review of *The Geological Evidences of the Antiquity of Man, Canadian Journal of Industry, Science, and Art*, n.s., 8 (1863): 389–394. See A. B. McKillop, *Matters of Mind: The University in Ontario, 1791–1951* (Toronto: University of Toronto Press, 1994), p. 119.

68. On the Hincks family and its accomplishments, see R. A. Baker, "The Reverend Thomas Hincks FRS (1818–1899) and His Family: Their Yorkshire Connections and Contributions to Natural History," *Naturalist* 124 (1999): 59–65; R. A. Baker, "The Hincks Dynasty: A Family of Unitarian Ministers and Their Contribution to Natural History and Education," in Nigel Cooper, ed., *John Ray and His Successors: The Clergyman as Biologist* (London: John Ray Trust, 1999), pp. 113–123. On William Hincks, see R. K. Webb, "Hincks, William," *Oxford Dictionary of National Biography* (Oxford: Oxford University Press, 2004), q.v.

69. McKillop, *A Disciplined Intelligence*, p. 113.

70. See the discussion in Jennifer Coggan, "Quinarianism after Darwin's *Origin*: The Circular System of William Hincks," *Journal of the History of Biology* 35 (2002): 5–42.

71. Coggan, "Quinarianism after Darwin's *Origin*," p. 10.

72. William Hincks, review of *On the Origin of Species, or the Causes of the Phenomena of Organic Nature: A Course of Lectures to Working Men*, by T. H. Huxley, *Canadian Journal of Industry, Science, and Art*, n.s., 8 (1863): 390–404, on pp. 390, 391, 393.

73. See Peter J. Bowler, "Darwinism and the Argument from Design: Suggestions for a Re-evaluation," *Journal of the History of Biology* 10 (1977): 29–43; David N. Livingstone, "The Idea of Design: The Vicissitudes of a Key Concept in the Princeton Response to Darwin," *Scottish Journal of Theology* 37 (1984): 329–357.

74. See Zeller, "Environment," p. 104.

75. R. Ramsay Wright, *An Introduction to Zoology. For the Use of High Schools* (Toronto: Copp, Clark Company, 1889), p. 269. Italics in original.

76. Wright, *Introduction to Zoology*, pp. 283–284, 285.

77. Wright, *Introduction to Zoology*, p. 285. On Cope, see Peter J. Bowler, "Edward Drinker Cope and the Changing Structure of Evolutionary Theory," *Isis* 68 (1977): 249–265; and Peter J. Bowler, *The Eclipse of Darwinism: Anti-Darwinian Evolution Theories in the Decades around 1900* (Baltimore: Johns Hopkins University Press, 1983). Bowler and Morus point out that in his *Theology of Evolution* of 1887, Cope argued that "the ability of the animals to direct evolution through their own efforts could be seen as God's creativity delegated into the life force that animates them." Peter J. Bowler and Iwan Rhys Morus, *Making Modern Science* (Chicago: University of Chicago Press, 2005), p. 358.

78. Berger, *Science, God and Nature*, p. 70.

79. Berger, *Science, God and Nature*, pp. 14, 15.

80. Zeller, *Inventing Canada*, p. 5.

81. Zeller, *Inventing Canada*, p. 4.

82. Brian J. Fraser, *Church, College, and Clergy: A History of Theological Education at Knox College Toronto, 1844–1994* (Montreal: McGill-Queen's University Press, 1995), p. 28. On the intellectual links between Canadian Presbyterianism and Scottish theology and philosophy, see also Michael Gauvreau, *The Evangelical Century: College and Creed in English Canada from the Great Revival to the Great Depression* (Montreal: McGill-Queen's University Press, 1991).

83. Nina Reid, "Christian Darwinism in the Knox College Monthly, 1883–1896," *Journal of the Canadian Church History Society* 31 (1989): 15–32.

84. A. C. Cheyne, *The Transforming of the Kirk: Victorian Scotland's Religious Revolution* (Edinburgh: St. Andrew Press, 1983), p. 174; Fraser, *Church, College, and Clergy*, p. 81.

85. Cited in Friedland, *University of Toronto*, p. 105.

86. G. M. Milligan, "The Religious Significance of the Mosaic Cosmogony," *Knox College Monthly* 4 (Apr. 1886): 271–281, on pp. 273–274.

87. The episode is discussed in David W. Bebbington, *The Mind of Gladstone: Religion, Homer and Politics* (Oxford: Oxford University Press, 2004), pp. 229–242.

88. Milligan, "Religious Significance of the Mosaic Cosmogony," p. 271.

89. Milligan, "Religious Significance of the Mosaic Cosmogony," p. 275.

90. John Hedley Brooke, "Darwin's Science and His Religion," in John Durant, ed., *Darwinism and Divinity: Essays on Evolution and Religious Belief* (Oxford: Blackwell, 1985), pp. 40–75.

91. Reid, "Christian Darwinism," p. 17.

92. W. A. Hunter, "Evolution and the Church," *Knox College Monthly* 13 (May 1895): 591–602, on p. 596.

93. See James R. Moore, *The Post-Darwinian Controversies: A Study of the Protestant Struggle to Come to Terms with Darwin in Great Britain and America, 1870–1900* (Cambridge: Cambridge University Press, 1979), pp. 273–275; Adrian Desmond and James Moore, *Darwin* (New York: Time Warner, 1991).

94. Hunter, "Evolution and the Church," p. 596.

95. Hunter, "Evolution and the Church," pp. 592, 596.

96. W. Dewar, "Biology and Theology," *Knox College Monthly* 4 (Feb. 1886), 150–156, on p. 153.

97. Dewar, "Biology and Theology," p. 155.

98. A. Blair, "Nature's Voice to Man's Religious Instincts," *Knox College Monthly* 3 (Feb. 1885): pp. 132–138, on p. 133.

99. Blair, "Nature's Voice," pp. 134–135.

100. F. R. Beattie, "The Design Argument: Its Scope and Import," *Knox College Monthly* 4 (Dec. 1885): 49–56, on pp. 54, 55. Also F. R. Beattie, "The Design Argument: Objections Considered," *Knox College Monthly* 4 (Feb. 1886): 141–150.

101. Hunter, "Evolution and the Church," p. 599, 598, 599.

102. Interestingly, it was Donald Harvey MacVicar, first principal of the Presbyterian College in Montreal, addressing the Knox College community in April 1893, who retained unstinting loyalty to the "inductive method" and urged that every new proposal should be rigorously tested "by the canons of induction." J. W. Dawson played a critical lay role in the early history of the college. See D. H. MacVicar, "Dogma and Current Thought," *Knox College Monthly* 11 (1893): 689–703, on p. 690.

103. Thomson, "Evolution in the Manifestation of the Supernatural," p. 301.

104. See Michael Gauvreau, "Baconianism, Darwinism, Fundamentalism: A Transatlantic Crisis of Faith," *Journal of Religious History* 13 (1985): 434–444.

105. See James Stalker, "The Present Desiderata of Theology," *Knox College Monthly* 8 (1890): 152–153.

106. Peter Galison, *Image and Logic: A Material Culture of Microphysics* (Chicago: University of Chicago Press, 1997).

107. See the discussion in John W. Rogerson, *Old Testament Criticism in the Nineteenth Century* (London: S.P.C.K, 1984).

108. R. Y. Thomson, "The Evolution in the Manifestation of the Supernatural," *Knox College Monthly* 12 (1890): 293–318, on p. 295.

109. The principle of evolutionary change was so deeply ingrained in the nineteenth-century documentary study of the Hebrew Bible that John Rogerson, excusing his lack of attention to the *Origin of Species* in his survey of nineteenth-century Old Testament scholarship, records that "one gets no indication of an almighty conflict occasioned by Darwin." John W. Rogerson, *Old Testament Criticism in the Nineteenth Century* (London: S.P.C.K., 1984), p. 253.

110. Thomson, "The Evolution in the Manifestation," p. 298.

111. Fraser, *Church, College, and Clergy,* p. 98.

112. J. Thompson, "Evolution of Scripture," *Knox College Monthly* 13 (1895): 363–372, on pp. 366, 363, 367.

113. Blair, "Nature's Voice," p. 136.

114. W. Caven, "Clerical Conservatism and Scientific Radicalism," *Knox College Monthly* 14 (1891): 285–295, on pp. 287, 295.

115. See Zeller, "Merchants of Light," p. 116.

116. " 'Theology and Religion': A Review of Dr. Caven's Lecture at the Opening of Knox College," *Queen's Journal* 5, no. 1 (20 Oct. 1877): 2.

117. Gauvreau, *Evangelical Century*, p. 148.

118. Berger, *Science, God, and Nature*.

119. J. Melbye and C. Meiklejohn, "A History of Physical Anthropology and the Development of Evolutionary Thought in Canada," *Human Evolution* 7, no. 3 (1992): 49–55, on p. 50.

Chapter 5 · Columbia, Woodrow, and the Legacy of the Lost Cause

1. *Complaint of James Woodrow versus The Synod of Georgia in Case of The Presbyterian Church in the United States (Rev. Dr. Wm. Adams, Voluntary Prosecutor) Versus James Woodrow: Argument of the Complainant, James Woodrow* (Columbia, SC: Presbyterian Publishing House, 1888), p. 12.

2. Andrew Dickson White, *A History of the Warfare of Science with Theology in Christendom*, vol. 1 (1896; New York: George Braziller, 1995), p. 318.

3. Andrew Dickson White, "New Chapters in the Warfare of Science. XIX. From Creation to Evolution. Part IV. The Final Effort of Theology," *Popular Science Monthly* 45 (June 1894): 145–159, on p. 145.

4. T. Watson Street, "The Evolution Controversy in the Southern Presbyterian Church with Attention to the Theological and Ecclesiastical Issues Raised," *Journal of the Presbyterian Historical Society* 37 (1959): 232–250, on p. 232.

5. "Woodrow's Heresy Trial. The Minister Defends the Theory of Evolution," *New York Times*, 23 May 1888.

6. Street, "The Evolution Controversy," p. 238.

7. "An Evolutionist Under Fire," *Springfield Daily Republican*, 28 May 1888, p. 4.

8. "An Evolutionist Under Fire," p. 4.

9. Quoted in Clement Eaton, "Professor James Woodrow and the Freedom of Teaching in the South," *Journal of Southern History* 28 (1962): 3–17, on p. 3.

10. R. A. Webb, "The Evolution Controversy," in George A. Blackburn, ed., *The Life Work of John L. Girardeau* (Columbia, SC: The State Company, 1916), pp. 231–284.

11. The whole episode is examined in detail in Frank J. Smith, "The Philosophy of Science in Late 19th Century Southern Presbyterianism" (Ph.D. diss., City University New York, 1992); see also Frank J. Smith, "Presbyterians and Evolution in the 19th Century: The Case of James Woodrow" *Contra Mundum* 6 (1993), 2–12; Ernest Trice Thompson, *Presbyterians in the South* (Richmond, VA: John Knox Press, 1973), vol. 2.

12. Street, "The Evolution Controversy," p. 237.

13. James Woodrow, *Evolution. An Address Delivered May 7th, 1884, Before the Alumni Association of the Columbia Theological Seminary* (Columbia, SC: Presbyterian Publishing House, 1884), p. 3. The lecture was also published in the July 1884 issue of the *Southern Presbyterian Review*.

14. For biographical details, see Robert K. Gustafson, *James Woodrow (1828–1907): Scientist, Theologian, Intellectual Leader* (Lewiston, NY: Edwin Mellen, 1995).

15. Woodrow, *Evolution*, p. 5, 6, 8.

16. Woodrow, *Evolution*, pp. 10, 11, 16.

17. I have discussed Mivart's stance in chapter 6 of David N. Livingstone, *Adam's Ancestors: Race, Religion and the Politics of Human Origins* (Baltimore: Johns Hopkins University Press, 2008. See also Jacob W. A. Gruber, *A Conscience in Conflict: The Life of St. George Jackson Mivart* (Westport, CT: Greenwood Press, 1960); Don O'Leary, *Roman Catholicism and Modern Science: A History* (New York: Continuum, 2006), chapter 4.

18. Woodrow, *Evolution*, pp. 25, 26, 28. It should be noted that Woodrow believed the Genesis text required the special creation of Eve. The significance of Eve in evolution-religion debates at the time is discussed in Diarmid Finnegan, "Eve and Evolution: Christian Evolutionists and the First Woman Question, 1860–1900," *Journal of the History of Ideas* (forthcoming). Here Finnegan comments: "Treating Eve in the same way as he had treated Adam would have added severely to his difficulties. Tampering with Eve's creation, even by religious conservatives, could too readily be tied to the cause of woman's rights, a movement roundly condemned by Southern Presbyterians."

19. "Columbia Seminary and Evolution," *Christian Observer* 63, no. 26 (June 25, 1884): 4.

20. Cited in Thompson, *Presbyterians in the South*, 2:466.

21. See Smith, "Presbyterians and Evolution."

22. Cited in Thompson, *Presbyterians in the South*, 2:466.

23. H. B. Pratt, "Evolution and Candidates for the Ministry," *Christian Observer* 63, no. 38 (Sept. 17, 1884): 4.

24. Quoted in Thompson, *Presbyterians in the South*, 2:465.

25. Eaton, "Professor James Woodrow," p. 10.

26. Quoted in "Professor Woodrow's Speech before the Synod of South Carolina," *Southern Presbyterian Review* 36 (1885): 1–65, on p. 5.

27. Synod of South Carolina, *Minutes of the Synod of South Carolina, at Its Annual Sessions at Greenville, S.C. October 22nd–28th, 1884* (Spartanburg, SC: T. J. Trimmier, 1884), p. 12, quoted in Smith, "Presbyterians and Evolution," p. 6.; Thompson, *Presbyterians in the South*, 2:469; "Professor Woodrow's Speech," p. 21.

28. Quoted in Thompson, *Presbyterians in the South*, 2:469, 470.

29. Street, "The Evolution Controversy," p. 241.

30. "Professor Woodrow's Speech," pp. 54, 65.

31. "Professor Woodrow's Speech," pp. 20, 12.

32. "Professor Woodrow's Speech," pp. 27, 17, 29, 18.

33. "Professor Woodrow's Speech," pp. 21, 25.

34. Blackburn, *The Life Work of John L. Girardeau.*

35. John L. Girardeau, *The Substance of Two Speeches on the Teaching of Evolution in Columbia Theological Seminary, Delivered in the Synod of South Carolina, at Greenville, S.C., Oct. 1884* (Columbia, SC: William Sloane, 1885), pp. 15, 10, 13.

36. Girardeau, *The Substance of Two Speeches*, pp. 23, 16, 17, 22, 27.

37. Girardeau, *The Substance of Two Speeches*, pp. 17, 18.

38. Girardeau, *The Substance of Two Speeches*, pp. 27, 29, 35.

39. John B. Adger's recollection of the entire controversy, with extensive quotations from the speeches delivered, can be found in John B. Adger, *My Life and Times, 1810–1899* (Richmond, VA: Presbyterian Committee, 1899), 412–666. The extracts cited in this paragraph come from pp. 462, 474, 477, 494.

40. The precise words—which Woodrow later quoted (somewhat to Girardeau's embarrassment, it seems) in his cross-examination of Girardeau during his trail before the Augusta Presbytery in 1886—were: "Insist Synod's action was no compromise; was definitely anti-Woodrow, so intended, so was"; quoted in Adger, *My Life and Times*, p. 564.

41. Adger, *My Life and Times*, pp. 526, 528–529, 531, 527.

42. Street, "The Evolution Controversy," p. 241; Thompson, *Presbyterians in the South*, 2:474.

43. Thompson, *Presbyterians in the South*, 2:477.

44. Geo. D. Armstrong, *Evolution. The Substance of Two Lectures* (Norfolk, VA: J. D. Ghiselin, 1885), p. 19.

45. J. William Flinn, "Evolution and Theology: The Logic of Prof. Woodrow's Opponents Examined," *Southern Presbyterian Review* 36 (1885): 268–304, on pp. 269, 270.

46. J. William Flinn, "Evolution and Theology: The Consensus of Science Against Dr. Woodrow's Opponents," *Southern Presbyterian Review* 36 (1885): 507–589, on pp. 564–565.

47. Later, when his application for membership of the Charleston Presbytery was declined in 1890, a lengthy list of such positions—presidencies of a home insurance company and national bank, vice presidencies of building associations, directorships of phosphate and furniture companies, and the like—was put to him as evidence of an "almost wholly secularized" life. *The Examination of the Rev. James Woodrow, D.D., by the Charleston Presbytery* (Charleston, SC: Lucas & Richardson Company, 1890).

48. *Springfield Republican*, Aug. 21, 1886.

49. Adger, *My Life and Times*, p. 648.

50. J. William Flinn, *An Argument on the Complaint of Rev. Jas. Woodrow, D.D. Against the Synod of Georgia* (n.p., n.d.), p. 10.

51. Gillespie, *A Defense of True Presbyterianism Against Two Deliverances of the Augusta Assembly, May, 1886* (n.p., n.d.), pp. 9, 14, 12, 13, 19.

52. The statement is available in several places: Adger, *My Life and Times*, p. 541; Thompson, *Presbyterians in the South*, 2:481.

53. *Complaint of James Woodrow versus The Synod of Georgia in Case of The Presbyterian Church in the United States versus James Woodrow* (Columbia, SC: Presbyterian Publishing House, 1888), p. 33.

54. *Complaint of James Woodrow versus The Synod of Georgia*, p. 34.

55. Gillespie, *A Defense of True Presbyterianism*, p. 11.

56. I have discussed this whole subject in Livingstone, *Adam's Ancestors*. On polygenism more generally, see William Stanton, *The Leopard's Spots: Scientific Attitudes toward Race in America, 1815–1859* (Chicago: University of Chicago Press, 1960).

57. Flinn, "Evolution and Theology," pp. 530, 525, 527.

58. Armstrong, *Evolution*, p. 7.

59. John Bachman, *The Doctrine of the Unity of the Human Race Examined on the Principles of Science* (Charleston, SC: C. Canning, 1850), pp. 212, 37, 287, 211.

60. Armstrong, *Evolution*, pp. 7, 20.

61. Adrian Desmond and James Moore, *Darwin's Sacred Cause: Race, Slavery and the Quest for Human Origins* (London: Allen Lane, 2009), pp. 210, 212. See also Lester D. Stephens, *Science, Race, and Religion in the American South: John Bachman and the Charleston Circle of Naturalists, 1815–1895* (Chapel Hill: University of North Carolina Press, 2000); Livingstone, *Adam's Ancestors*, chapter 7.

62. Flinn, "Evolution and Theology," p. 523. Emphasis in original.

63. Flinn, "Evolution and Theology," p. 575.

64. Flinn, "Evolution and Theology," p. 509.

65. The resolution is reprinted in Eaton, "Professor James Woodrow," p. 4; White, *History of the Warfare*, 1:316; James A. Lyon, "The New Theological Professorship—Natural Science in Connexion with Revealed Religion," *Southern Presbyterian Review* 12 (1859): 181–195, on p. 182; James Woodrow, "Inaugural Address," in Marion W. Woodrow, ed., *Dr James Woodrow as Seen by his Friends: Character Sketches by His Former Pupils, Colleagues, and Associates* (Columbia, SC: R. L. Bryan, 1909), pp. 365–387, on pp. 367–368.

66. Lyon, "The New Theological Professorship," pp. 183, 184, 186.

67. Richard S. Gladney, "Natural Science and Revealed Religion," *Southern Presbyterian Review* 12 (1859): 443–467, on pp. 458, 459, 444, 447.

68. Details are recorded in Gustafson, *James Woodrow*, chapter 3.

69. See Sean Michael Lucas, *Robert Lewis Dabney: A Southern Presbyterian Life* (Phillipsburg, NJ: P&R Publishing, 2005).

70. Quoted in Adger, *My Life and Times*, p. 424.

71. Thomas Smyth, *Autobiographical Notes, Letters and Reflections* (Charleston: Walker, Evans and Cogswell Co., 1914), p. 639.

72. [Robert Dabney], "Geology and the Bible," *Southern Presbyterian Review* 14 (July 1861): 246–274, on pp. 247, 249.

73. In essence, this was the idea that had been promulgated by Philip Henry Gosse in *Omphalos*, namely, prochronism or the creation of a universe with the appearance of age. See Ann Thwaite, *Glimpses of the Wonderful: The Life of Philip Henry Gosse, 1810–1888* (London: Faber & Faber, 2002). Later, in 1863, Dabney, recording the "glow" that this "new discovery" had induced, observed that "Mr. P.H. Gosse, a British naturalist, advanced substantially the same idea in a book quaintly called, *Omphalos*." Robert L. Dabney, "The Caution Against Anti-Christian Science Criticised by Dr. Woodrow," *Southern Presbyterian Review* 24 (1873): 539–585. Reprinted in C. R. Vaughan, *Discussions by Robert L. Dabney*, 4 vols. (Richmond, VA: Presbyterian Committee of Publication, 1890–97), 3:137–180, on p. 170.

74. [Dabney], "Geology and the Bible," pp. 265, 253.

75. Woodrow, "Inaugural Address," pp. 377, 379, 383–384, 386.

76. [James Woodrow], "Geology and Its Assailants," *Southern Presbyterian Review* 15 (1863): 549–568, reprinted in Marion W. Woodrow, *Dr James Woodrow*, pp. 388–406.

77. Woodrow, "Geology and Its Assailants," p. 395; [Dabney], "Geology and the Bible," p. 248.

78. Woodrow, "Geology and Its Assailants," p. 392.

79. J. R. Blake, "The Geological Writings of David N. Lord," *Southern Presbyterian Review* 13 (1861): 537–577. On "scriptural geology," see Michael Roberts, "Geology and Genesis Unearthed," *Churchman* 112 (1998): 225–255; Ronald L. Numbers, *The Creationists* (New York: Alfred A. Knopf, 1992); Rodney L. Stiling, "Scriptural Geology in America," in David N. Livingstone, D. G. Hart, and Mark A. Noll, eds., *Evangelicals and Science in Historical Perspective* (Oxford: Oxford University Press, 1999), pp. 177–192; Ralph O'Connor, "Young-Earth Creationists in Early Nineteenth Century Britain? Towards a Reassessment of 'Scriptural Geology,'" *History of Science* 45 (2007): 357–403.

80. Woodrow, "Geology and Its Assailants," p. 395.

81. Robert L. Dabney, "Memorial on Theological Education," in Vaughan, *Discussions,* 2:47–75, on pp. 72, 73.

82. Robert L. Dabney, "A Caution Against Anti-Christian Science," in Vaughan, *Discussions,* 3:116–136, on pp. 117–118 119, 126.

83. Dabney, "Memorial on Theological Education," p. 73.

84. Dabney, "A Caution Against Anti-Christian Science," pp. 121, 123.

85. James Woodrow, "An Examination of Certain Recent Assaults on Physical Science," *Southern Presbyterian Review* 24 (July 1873): 327–376, on pp. 328, 329. The piece is reprinted in Marion W. Woodrow, *Dr James Woodrow,* pp. 407–459.

86. Woodrow, "An Examination of Certain Recent Assaults on Physical Science," pp. 9, 10, 11, 12, 15.

87. Woodrow, "An Examination of Certain Recent Assaults on Physical Science," p. 27.

88. Dabney, "The Caution Against Anti-Christian Science Criticised by Dr. Woodrow," pp. 139, 148 141, 155, 142, 145.

89. James Woodrow, "A Further Examination of Certain Recent Assaults on Physical Science," *Southern Presbyterian Review* 25 (Apr. 1874): 246–291, on pp. 253, 290–291. Reprinted in Marion W. Woodrow, *Dr James Woodrow,* pp. 460–507.

90. R. T. Brumby, "Relations of Science to the Bible," *Southern Presbyterian Review* 25 (1874): 1–31, on pp. 3, 2, 3, 4.

91. R. T. Brumby, "Gradualness Characteristic of all God's Operations," *Southern Presbyterian Review* 25 (1874): 524–555, on pp. 540, 547.

92. W. S. Bean, "The Outlook of Modern Science," *Southern Presbyterian Review* 26 (1874): 331–338, on pp. 331, 333, 334.

93. Robert L. Dabney, *The Sensualistic Philosophy of the Nineteenth Century Considered* (New York: Anson D. F. Randolph, 1876), pp. 110, 114, 128.

94. Dabney, *Sensualistic Philosophy,* pp. 175, 176, 177, 181, 187.

95. Dabney, *Sensualistic Philosophy,* p. 206.

96. On Dabney and the Lost Cause, see Eugene D. Genovese, *A Consuming Fire: The Fall of the Confederacy in the Mind of the White Christian South* (Athens: University of Georgia Press, 1998); Eugene D. Genovese, *The Southern Tradition: The Achievements and Limitations of an American Conservatism* (Cambridge, MA: Harvard University Press, 1994); Lucas, *Robert Lewis Dabney;* Monte Hampton, "Navigating Modernity: The Bible, the New South, and Robert Lewis Dabney," in Bertram Wyatt-Brown and Peter Wallentstein, eds., *Virginia's Civil War* (Charlottesville: University of Virginia Press, 2004), pp. 216–229; Monte Harrell Hampton, " 'Handmaid' or 'Assailant': Debating Science and Scripture in the Culture of the Lost Cause" (Ph.D. diss., University of North Carolina at Chapel Hill, 2004); Hampton's revised dissertation is forthcoming as *Storm of Words: Science, Religion, and Southern Culture in the Era of the American Civil War* (Tuscaloosa: University of Alabama Press); Gaines M. Foster, *Ghosts of the Confederacy: Defeat, the Lost Cause, and the Emergence of the New South, 1865 to 1913* (New York: Oxford University Press, 1987).

97. Robert L. Dabney, "The Negro and the Common School," *Southern Planter and Farmer* 37 (Apr. 1876): 251–262, on pp. 251, 253.

98. On the history and culture of hybridity, see Robert J. C. Young, *Colonial Desire: Hybridity in Theory, Culture and Race* (London: Routledge, 1995).

99. Dabney, "The Negro and the Common School," pp. 252, 253, 258, 259.

100. Quoted in William D. Carrigan, "In Defense of the Social Order: Racial Thought among Southern Presbyterians in the Nineteenth Century," *American Nineteenth Century History* 1 (2000): 31–52, on p. 39.

101. Dabney, The Negro and the Common School," p. 253.

102. Dabney's side of the correspondence is collected under the running title "The State Free School System Imposed Upon Virginia by the Underwood Constitution," in Vaughan, *Discussions*, 4:238–271.

103. Dabney, "The State Free School System," pp. 240, 241, 247, 249, 255.

104. Dabney had earlier defended slavery on biblical grounds in a series of articles, "The Moral Characteristics of Slavery," for the *Richmond Inquirer* in 1851. See Carrigan, "In Defense of the Social Order."

105. Robert L. Dabney, *A Defence of Virginia: (and Through Her, of the South) in Recent and Pending Contests against the Sectional Party* (New York: E. J. Hale, 1867), pp. 6, 21, 25–26.

106. On the role of paternalism, patriarchy, and reciprocal duties in the southern slave system, see Eugene Genovese, *Roll, Jordan, Roll: The World the Slaves Made* (New York: Pantheon, 1972); Elizabeth Fox-Genovese and Eugene Genovese, *The Mind of the Master Class: History in the Southern Slaveholders' Worldview* (Cambridge: Cambridge University Press, 2005).

107. Dabney, *Defence of Virginia*, pp. 22–23.

108. Dabney, *Defence of Virginia*, p. 353.

109. On this mindset more generally, see H. Sheldon Smith, *In His Image, But . . . Racism in Southern Religion, 1780–1910* (Durham, NC: Duke University Press, 1972).

110. See Carrigan, "In Defense of the Social Order"; Mark A. Noll, *America's God: From Jonathan Edwards to Abraham Lincoln* (New York: Oxford University Press, 2002), chapter 19, "The Bible and Slavery"; Mark A. Noll, *The Civil War as a Theological Crisis* (Chapel Hill: University of North Carolina Press, 2006); Christopher A. Luse, "Slavery's Champions Stood at Odds: Polygenesis and the Defense of Slavery," *Civil War History* 53 (2007): 379–412.

111. Thomas Smyth, *The Unity of the Human Races Proved to be the Doctrine of Scripture, Reason and Science with a Review of the Present Position and Theory of Professor Agassiz* (New York: Putnam, 1850), p. 333–334.

112. Smyth, *Unity of the Human Races*, p. 337.

113. George Howe, review of "Two Lectures on the Biblical and Physical History of Man . . . by Josiah C. Nott," *Southern Presbyterian Review* 3 (1850): 426–490, on pp. 429, 427, 473, 489, 486, 487.

114. Robert Lewis Dabney, "Anti-Biblical Theories of Rights," *Presbyterian Quarterly* 2 (July 1888): 217–242, on pp. 217, 218, 219, 222, 223, 231, 240.

115. Ronald L. Numbers, *Darwinism Comes to America* (Cambridge, MA: Harvard University Press, 1998), p. 65. See also Ronald L. Numbers and Lester D. Stephens, "Darwinism in the American South," in Ronald L. Numbers and John Stenhouse, eds., *Disseminating Darwinism: The Role of Place, Race, Religion, and Gender* (Cambridge: Cambridge University Press, 1999), pp. 123–143.

116. George D. Armstrong, *The Two Books of Nature and Revelation Collated* (New York: Funk & Wagnalls, 1886), p. 95.

117. Armstrong, *Two Books of Nature*, pp. 181, 183.

118. Geo. D. Armstrong, *The Christian Doctrine of Slavery* (New York: Scribner, 1857), pp. iii, 100, 131–132. In large measure, this volume was an extended critique of Albert Barnes, *The Church and Slavery* (Philadelphia: Parry & McMillan, 1857).

119. See Hampton, "Handmaid."

120. Quoted in Thompson, *Presbyterians in the South*, 2:450.

121. Girardeau, *The Substance of Two Speeches*, p. 25.

122. J. L. Girardeau, "Appendix. Reply to Dr. Martin," appended to James L. Martin, *Dr. Girardeau's Anti-Evolution: The Logic of His Reply* (Columbia, SC: Presbyterian Printing House, 1889), p. 63.

123. Martin, *Dr. Girardeau's Anti-Evolution*, p. 54.

124. Ariel [Buckner H. Payne], *The Negro: What is His Ethnological Status? Is He the Progeny of Ham? Is He a Descendant of Adam and Eve? Has He a Soul? Or Is He a Beast in God's Nomenclature? What is His Status as Fixed by God in Creation? What is His Relation to the White Race?* (Cincinnati: Publisher for the Proprietor, 1867), p. 22.

125. I have discussed this whole episode in Livingstone, *Adam's Ancestors*. See also Colin Kidd, *The Forging of Races: Race and Scripture in the Protestant Atlantic World, 1600–2000* (Cambridge: Cambridge University Press, 2006), pp. 149–150; George M. Frederickson, *The Black Image in the White Mind: The Debate on Afro-American Character and Destiny, 1817–1914* (New York: Harper and Row, 1972), pp. 188–189; Mason Stokes, "Someone's in the Garden with Eve: Race, Religion, and the American Fall," *American Quarterly* 50 (1998): 718–743; Stephen R. Haynes, *Noah's Curse: The Biblical Justification of American Slavery* (Oxford: Oxford University Press, 2002), pp. 112–113.

126. J. L. Girardeau, "Is the Negro a Beast? A Discourse delivered by Rev. J. L. Girardeau, D.D, in Zion Presbyterian Church, Calhoun St., Sunday Evening, November 17, 1867," *Charleston Mercury*, 19 Nov. 1867.

127. Cited in Leonard Alberstadt, "Alexander Winchell's Preadamites—A Case for Dismissal from the Vanderbilt University," *Earth Sciences History* 13 (1994): 97–112, on p. 110. I have discussed this whole episode in Livingstone, *Adam's Ancestors*, pp. 141–153.

128. Howe, review of "Two Lectures," p. 427, 489, 485.

129. James Woodrow, "Politics and Religion," in Marion W. Woodrow, *Dr James Woodrow*, pp. 601–603, on p. 603.

130. Woodrow, "Dr Palmer's Open Letter," in Marion W. Woodrow, *Dr James Woodrow*, pp. 582–586, on pp. 583–584. Woodrow's racial views are discussed in Gustafson, *James Woodrow*, p. 97 ff.

131. Woodrow, "Northern Ideas about Organic Union," in Marion W. Woodrow, *Dr James Woodrow*, pp. 590–92, on 591.

132. James Woodrow, "The Presbyterian Doctrine of the Bible," in Marion W. Woodrow, *Dr James Woodrow*, pp. 252–275, on p. 265. This was the text of an address he delivered in August 1886.

133. Geo. D. Armstrong, "The Word of God versus 'The Bible of Modern Scientific Theology,'" *Presbyterian Quarterly* 2 (1888): 39–58.

134. Carrigan, "In Defense of the Social Order," p. 41.

135. John B. Adger, "The Future of the Freedmen," *Southern Presbyterian Review* 19 (1868): 268–294, on p. 279.

136. John B. Adger, *Christian Missions and African Colonization* (Columbia, SC: Steam Power Press, 1857), p. 18.

Chapter 6 · *Princeton, Darwinism, and the Shorthorn Cattle*

1. The journal had been established in 1867 and would shortly move its entire operation to Louisville.

2. Ethelbert D. Warfield, "Biographical Sketch of Benjamin Breckinridge Warfield," in Ethelbert D. Warfield, William Park Armstrong, and Caspar Wistar Hodge, eds., *The Works of Benjamin B. Warfield*, 10 vols. (New York: Oxford University Press, 1932), 1:vi.

3. This scrapbook is extant in the Speer Library, Princeton Theological Seminary. Most of the clippings are from the *National Livestock Journal*. The Speer Library also retains a number of books on shorthorn cattle from B. B. Warfield's personal library.

4. "Miller, Dryden, and Warfield," *American Breeders Magazine* 2, no. 1 (Oct. 1911): 3–9.

5. John M. Gresham, "William Warfield," *Biographical Cyclopedia of the Commonwealth of Kentucky Embracing Biographies of Many of the Prominent Men and Families of the State* (Chicago: John M. Gresham Company, 1896), pp. 525–536.

6. William Warfield, *American Short-Horn Importations Containing the Pedigrees of all Short-Horn Cattle Hitherto Imported into America* (Chicago: American Breeders' Association, 1884); William Warfield, *The Theory and Practice of Cattle-Breeding* (Chicago: J. H. Sanders Publishing Company, 1889).

7. Warfield, *The Theory and Practice of Cattle-Breeding*, preface.

8. Warfield, *The Theory and Practice of Cattle-Breeding*, pp. 85–86. William Warfield also enthusiastically commended a volume by J. H. Sanders, *Horse-Breeding: Being the General Principles of Heredity applied to the Business of Breeding Horses . . .* (Chicago: J. H. Sanders, 1885), in which Darwin's evolutionary thinking was promulgated. Warfield particularly endorsed the long first chapter, which dealt with these general principles.

9. Warfield, *The Theory and Practice of Cattle-Breeding*, p. 70.

10. See James Secord, "Nature's Fancy: Charles Darwin and the Breeding of Pigeons," *Isis* 72 (1981): 163–186.

11. Warfield, *The Theory and Practice of Cattle-Breeding*, pp. 16–17.

12. Benjamin B. Warfield, "Personal Recollections of Princeton Undergraduate Life. IV—The Coming of Dr. McCosh," *Princeton Alumni Weekly* 16, no. 28 (19 Apr. 1916): 650–653, on p. 652.

13. See the discussion in Peter J. Bowler, *The Eclipse of Darwinism: Anti-Darwinian Evolution Theories in the Decades around 1900* (Baltimore: Johns Hopkins University Press, 1983).

14. These volumes, signed by Warfield and dated, are held in the Library of the Princeton Theological Seminary, Princeton, New Jersey.

15. Warfield, "Personal Recollections," p. 652.

16. Charles Hodge, *What is Darwinism?* (New York: Scribner, Armstrong & Co, 1874), p. 177.

17. See Bradley John Gundlach, "The Evolution Question at Princeton, 1845–1929" (Ph.D. diss., University of Rochester, New York, 1995).

18. For example, Robert Blakey, *The Angler's Complete Guide to the Rivers and Lakes of England* (Liverpool: Whittaker & Co., 1853); *Hints on Angling: With Suggestions for Angling Excursions in France and Belgium* (London: W. W. Robinson, 1846); *The Angler's Guide to the Rivers and Lochs of Scotland* (Glasgow: Thomas Murray, 1854).

19. Such as Robert Blakey, *History of Moral Science*, 2 vols. (London: James Duncan,

1833, 1836); *History of the Philosophy of Mind* (London: Longman, Brown, Green & Longmans, 1850); *The History of Political Literature from the Earliest Times* (London: Richard Bentley, 1855).

20. On Blakey, see Henry Miller, ed., *Memoirs of Robert Blakey: Professor of Logic and Metaphysics, Queen's College, Belfast* (London: Trübner, 1879), and Frederic Boase, *Modern English Biography: Containing Many Thousand Concise Memoirs of Persons who have Died since the Year 1850, with an Index of the Most Interesting Matter* (London: Frank Cass, 1965, orig. 1892–1921), q.v. Cousin's significance for English-speaking philosophy is the subject of George Elder Davie, "Victor Cousin and the Scottish Philosophers," *Edinburgh Review* 74 (1986), reprinted in Davie's *A Passion for Ideas: Essays on the Scottish Enlightenment* (Edinburgh: Polygon, 1994), 2:70–109.

21. The major biography is J. David Hoeveler Jr., *James McCosh and the Scottish Intellectual Tradition: From Glasgow to Princeton* (Princeton: Princeton University Press, 1981). The quotation comes from p. 96. See also William Milligan Sloane, ed., *The Life of James McCosh: A Record Chiefly Autobiographical* (Edinburgh: T. & T. Clark, 1896).

22. Thomas Reid, *Essays on the Intellectual Powers of Man* (1785; Cambridge, MA: MIT Press, 1969), p. 129.

23. See Nicholas Wolterstorff, *Thomas Reid and the Story of Epistemology* (Cambridge: Cambridge University Press, 2001); Heiner F. Klemme, "Scepticism and Common Sense," in Alexander Broadie, ed., *The Cambridge Companion to the Scottish Enlightenment* (Cambridge: Cambridge University Press, 2003), pp. 117–135.

24. See, for example, James McCosh, "Morphological Analogy between the Disposition of the Branches of Exogenous Plants and the Venation of their Leaves," *Report of the 22nd Meeting of the British Association for the Advancement of Science, held at Belfast* (1852): pp. 66–68; and "Some Observations on the Morphology of Pines and Firs," *Report of the 24th Meeting of the British Association for the Advancement of Science, held at Liverpool* (1854), pp. 99–100.

25. On transcendental morphology in general and Owen's work in particular, see Philip F. Rehbock, *The Philosophical Naturalists: Themes in Early Nineteenth-Century British Biology* (Madison: University of Wisconsin Press, 1983); Nicolaas A. Rupke, *Richard Owen, Biology without Darwin, a Revised Edition* (Chicago: University of Chicago Press, 2009).

26. This is the term used by Bowler. See Peter J. Bowler, "Darwinism and the Argument from Design: Suggestions for a Re-evaluation," *Journal of the History of Biology* 10 (1977): 29–43.

27. See Adrian Desmond, *The Politics of Evolution: Morphology, Medicine, and Reform in Radical London* (Chicago: University of Chicago Press, 1989).

28. I have discussed the issue of design among the Princeton thinkers more generally in David N. Livingstone, "The Idea of Design: The Vicissitudes of a Key Concept in the Princeton Response to Darwin," *Scottish Journal of Theology* 37 (1984): 329–357.

29. James McCosh and George Dickie, *Typical Forms and Special Ends in Creation*, 2nd ed. (Edinburgh: Thomas Constable, 1857), pp. 7, 8, 15, 27, 442.

30. McCosh and Dickie, *Typical Forms*, p. 40.

31. See, e.g., James R. Moore, *The Post-Darwinian Controversies: A Study of the Protestant Struggle to Come to Terms with Darwin in Great Britain and America, 1870–1900* (Cambridge: Cambridge University Press, 1979), pp. 245–251; David N. Livingstone, *Darwin's Forgotten*

Defenders: The Encounter between Evangelical Theology and Evolutionary Thought (Edinburgh: Scottish Academic Press, 1987).

32. *Inauguration of Rev. Jas. McCosh, D.D., LL.D., as President of Princeton College, October 27, 1868* (Princeton, NJ: Stelle and Smith, 1868), pp. 3, 5.

33. James McCosh, *Christianity and Positivism: A Series of Lectures to the Times on Natural Theology and Christian Apologetics* (London: Macmillan, 1871), pp. 42, 63.

34. McCosh, *Christianity and Positivism*, pp. 64, 66.

35. McCosh, *Christianity and Positivism*, pp. 346, 37, 351.

36. I have discussed polygenism in David N. Livingstone, *Adam's Ancestors: Race, Religion, and the Politics of Human Origins* (Baltimore: Johns Hopkins University Press, 2008).

37. Charles Hodge, "Examination of some Reasonings against the Unity of Mankind," *Biblical Repertory and Princeton Review* 34 (1862): 435–464, on p. 461.

38. Charles Hodge, *Systematic Theology*, vol. 2 (New York: Charles Scribner's, 1872), pp. 12, 16, 19, 21.

39. James McCosh, "Religious Aspects of the Doctrine of Development," in S. Irenaeus Prime, ed., *History, Orations, and Other Documents of the Sixth General Conference of the Evangelical Alliance* (New York: Harper, 1874), pp. 264–271, on pp. 266, 267, 270.

40. It turned out to be a controversial performance, as members of other denominations were troubled by some of his assumptions. See Prime, *History, Orations*, p. 144.

41. "Discussion on Darwinism and the Doctrine of Development," in Prime, ed., *History, Orations*, pp. 317–323, on pp. 318, 320.

42. Charles Hodge, *What is Darwinism?* (New York: Scribner's, 1874), pp. 41, 48, 108, 52.

43. Hodge, *What is Darwinism?* p. 174.

44. James McCosh, "On Evolution," in J. G. Wood, *Wood's Bible Animals. A Description of the Habits, Structure, and Uses of Every Living Creature Mentioned in the Scriptures, From the Ape to the Coral; and Explaining all Those Passages in the Old and New Testaments in Which Reference is Made to Beast, Bird, Reptile, Fish, or Insect*, American ed. (Philadelphia: Bradley, Garretson & Co., 1875), pp. 649–677, on pp. 649, 655, 659.

45. McCosh, "On Evolution," p. 660.

46. James McCosh, *Ideas in Nature Overlooked by Dr. Tyndall. Being an Examination of Dr. Tyndall's Belfast Address* (New York: Robert Carter and Brothers, 1875), pp. iv, 4.

47. McCosh, *Ideas in Nature*, pp. 15, 16, 33.

48. James McCosh, *The Religious Aspect of Evolution* (New York: Putnam, 1888), pp. ix, 7, 8, 16, 27.

49. McCosh, *Religious Aspect*, pp. 58–59.

50. McCosh, *Religious Aspect*, pp. 61–62, 68.

51. See chapter 2.

52. McCosh, *Religious Aspect*, p. 77.

53. McCosh, *Religious Aspect*, p. 89. In the light of these intimations, it is hardly surprising that Ernst Benz sensed reverberations between McCosh's evolutionary eschatology and Teilhard de Chardin's Omega point. See Ernst Benz, *Evolution and Christian Hope: Man's Concept of the Future from the Early Fathers to Teilhard de Chardin*, trans. from the German by Heinz G. Frank (New York: Doubleday, 1966). I have discussed this in David N. Livingstone, "Evolution and Eschatology," *Themelios* 22 (1996): 26–36.

54. *Inauguration of Rev. Jas. McCosh, D.D., LL.D., as President of Princeton College, October 27, 1868* (Princeton, NJ: Stelle and Smith, 1868), p. 6.

55. *Proceedings Connected with the Semi-centennial Commemoration of the Professorship of Rev. Charles Hodge, D.D., LL.D in the Theological Seminary at Princeton, N. J., April 24, 1872* (New York: Randolph, 1872), pp. 61, 64.

56. *Proceedings Connected with the Semi-centennial Commemoration*, pp. 60, 59.

57. "Evolution and Christianity," *New York Times*, 20 Apr. 1890.

58. Sloane, *The Life of James McCosh*, pp. 123, 124, 234–235.

59. See the discussion in chapter 3, and also James McCosh, *The Ulster Revival and Its Physiological Accidents. A Paper Read Before the Evangelical Alliance, September 22, 1859* (Belfast: Aitchison, 1859).

60. At the same time, he was judged by some back in Edinburgh to be too conservative. See Diarmid Finnegan, "Placing Science in an Age of Oratory: Spaces of Scientific Speech in Mid-Victorian Edinburgh," in David N. Livingstone and Charles W. J. Withers, eds., *Geographies of Nineteenth Century Science* (Chicago: University of Chicago Press, 2011), pp. 153–177.

61. George M. Marsden, *The Soul of the American University: From Protestant Establishment to Established Nonbelief* (New York: Oxford University Press, 1994), p. 204.

62. Sloane, *The Life of James McCosh*, p. 234.

63. McCosh, *Religious Aspect*, p. xi.

64. Marsden, *Soul of the American University*, p. 203. Gundlach also argues for the similarity of their evaluations. While they disagreed on the alleged fact of species transmutation and on the design issue, Gundlach claims, "they agreed on nearly everything else connected with the question of evolution and religion." See Bradley J. Gundlach, "McCosh and Hodge on Evolution: A Combined Legacy," *Journal of Presbyterian History* 75 (1997): 85–102, on p. 88.

65. *Proceedings Connected with the Semi-centennial Commemoration*, p. 52.

66. See the reassessments in John W. Stewart and James H. Moorehead, eds., *Charles Hodge Revisited: A Critical Appraisal of His Life and Work* (Grand Rapids, MI: Eerdmans, 2002).

67. Asa Gray, review of *What is Darwinism?* in *Nation*, 28 May 1874, pp. 348–351, on p. 351.

68. See James D. Dana, "Biographical Memoir of Arnold Guyot, 1807–1884," *Biographical Memoirs of the National Academy of Sciences* 2 (1886): 309–347. Also William Libbey Jr., "The Life and Scientific Work of Arnold Guyot," *Journal of the American Geographical Society* 16 (1884): 194–221; Edith H. Ferrell, "Arnold Henry Guyot, 1807–1884," in T. W. Freeman, ed., *Geographers Bibliographical Studies* 5 (1981): 63–71.

69. Arnold Guyot, *The Earth and Man: Lectures on Comparative Physical Geography, in its Relation to the History of Mankind* (Boston: Gould, Kendall and Lincoln, 1849).

70. I have discussed Guyot's geographical contributions in *The Geographical Tradition: Episodes in the History of a Contested Enterprise* (Oxford: Blackwell, 1992).

71. Arnold Guyot, "Cosmogony and the Bible; or, the Biblical Account of Creation in the Light of Modern Science," in Prime, *History, Orations*, pp. 276–287, on pp. 276, 286.

72. See the discussion of Guyot in this connection in Ronald L. Numbers, *Creation by Natural Law: Laplace's Nebular Hypothesis in American Thought* (Seattle: University of Washington Press, 1977), pp. 91–100.

73. For the quotation and further discussion, see Ronald L. Numbers, *The Creationists: The Evolution of Scientific Creationism* (New York: Alfred A. Knopf, 1992), p. 7.

74. Dana, "Biographical Memoir," p. 328.

75. This is also Numbers's view. See *Creationists*, p. 9.

76. Dana, "Biographical Memoir," p. 334.

77. Arnold Guyot, *Creation or the Biblical Cosmology in the Light of Modern Science* (New York: Scribner's, 1884), pp. 127, 128.

78. See the discussion and the cited extract in Susan Sheets-Pyenson, *John William Dawson: Faith, Hope, and Science* (Montreal: McGill-Queen's University Press, 1996), pp. 87–90.

79. Quoted in Gundlach, "Evolution Question at Princeton," p. 174.

80. For Cope's metaphysical evolutionism, including his biological approach to consciousness and the origin of the will, see Edward Drinker Cope, *The Origin of the Fittest: Essays on Evolution* (New York: Appleton, 1887).

81. Details about the post are discussed in Gundlach, "Evolution Question at Princeton," 167–171. See also Edwin G. Conklin, "Biology at Princeton," *Bios* 19 (1948): 151–171.

82. James McCosh to Rev. George Macloskie, 29 Aug. 1871, McCosh Correspondence, General Mss (misc) CO140, Firestone Library, Princeton University.

83. I have discussed Macloskie in Livingstone, *Darwin's Forgotten Defenders*, pp. 92–96; and in David N. Livingstone and Ronald A. Wells, *Ulster-American Religion: Episodes in the History of a Cultural Connection* (Notre Dame: University of Notre Dame Press, 1999), pp. 40–48.

84. This is recorded in a manuscript by George Macloskie entitled "Biographical Sketch," housed in the Macloskie Papers, CO498, Carton 4, Firestone Library, Princeton University.

85. Quoted in Conklin, "Biology at Princeton," p. 157.

86. These letters are extant in the Macloskie Papers, CO498, Carton 4, Firestone Library, Princeton University.

87. G. Macloskie, "The Natural History of Man," *Ulster Magazine* 3 (1863): 217–237, on p. 230. Prior to leaving Ireland, he had published "The Silicified Wood of Lough Neagh, with Notes on the Structure of Coniferous Wood," *Proceedings of the Belfast Natural History and Philosophical Society,* 14 Feb. 1872, pp. 51–74.

88. Miscellaneous Zoology Notes, n.d., Macloskie Papers, CO498, Box 1, Firestone Library, Princeton University.

89. This was notwithstanding his personal friendship with the Irish theologian Robert Watts, who strenuously opposed evolution. Watts's anti-Darwinism is discussed in chapter 3. When Watts visited Princeton in 1880, he wrote to his wife: "After writing you at Professor Green's on Monday, Dr. M'Closkie who you remember lost his dredge in Belfast Lough when our little boys were out with him in a boat, called according to promise and took me over all the buildings of Dr M'Cosh's College." Letter, Robert Watts, West Philadelphia, to His Wife, 17 Sept. 1880, in "Family Letters of Revd. Robert Watts, D.D., LL.D.," compiled by his Wife (Typescript). I am grateful to Dr. R. E. L. Rodgers for making this typescript available to me.

90. "Biographical Sketch," Macloskie Papers, CO498, Carton 4, Firestone Library, Princeton University.

91. George Macloskie, "Scientific Speculation," *Presbyterian Review* 8 (1887): 617–625, on p. 618.

92. George Macloskie, "Concessions to Science," *Presbyterian Review* 10 (1889): 220–228, on p. 222.

93. Macloskie, "Scientific Speculation," p. 620.

94. Macloskie, "Concessions to Science," p. 227.

95. His papers contain the detailed record of the action taken against Woodrow. These include *The Examination of the Rev. James Woodrow, D.D., by the Charleston Presbytery* (Charleston, SC: Lucas & Richardson Co., 1890).

96. George Macloskie, "Common Errors as to the Relations of Science and Faith," *Presbyterian and Reformed Review* 6 (1895): 98–107, on pp. 99, 104.

97. George Macloskie, review of *Life and Letters of Charles Darwin*, by Francis Darwin, *Presbyterian Review* 9 (1888): 519–522, on p. 522.

98. George Macloskie, "Theistic Evolution," *Presbyterian and Reformed Review* 33 (1898): 1–22, on p. 21.

99. George Macloskie, "The Origin of New Species and of Man," *Bibliotheca Sacra* 60 (1903): 261–275, on p. 273.

100. George Macloskie, "The Drift of Modern Science," ms. in notebook, Macloskie Papers, CO498, Carton 1, Firestone Library, Princeton University.

101. Mss Notebook by George Macloskie, "Drift of Modern Science," Macloskie Papers, CO498, Carton 1, Princeton University Library.

102. George Macloskie, "The Outlook of Science and Faith," *Princeton Theological Review* 1 (1903): 597–615, on p. 602.

103. William Berryman Scott, *Some Memories of a Palaeontologist* (Princeton: Princeton University Press, 1939), pp. 49, 75.

104. G. G. Simpson, "Biographical Memoir of William Berryman Scott, 1858–1947," *Biographical Memoirs of the National Academy of Sciences* 25 (1948): 173–203, on p. 188.

105. Scott, *Some Memories of a Palaeontologist*, p. 223.

106. See Peter J. Bowler, *The Eclipse of Darwinism: Anti-Darwinian Evolution Theories in the Decades around 1900* (Baltimore: Johns Hopkins University Press, 1983).

107. Simpson, "Biographical Memoir of William Berryman Scott," pp. 189, 188, 191.

108. William Berryman Scott, *The Theory of Evolution: With Special Reference to the Evidence upon Which it is Founded* (New York: The Macmillan Company, 1917), p. v.

109. Scott, *Theory of Evolution*, p. 25.

110. Osborn's career at the American Museum of Natural History is examined in Ronald Rainger, *An Agenda for Antiquity: Henry Fairfield Osborn and Vertebrate Paleontology at the American Museum of Natural History, 1890–1935* (Tuscaloosa: University of Alabama Press, 1991); his thinking about race and evolution are the subject of Brian Regal, *Henry Fairfield Osborn: Race and the Search for the Origins of Man* (London: Ashgate, 2002).

111. William K. Gregory, "Biographical Memoir of Henry Fairfield Osborn, 1857–1935," *Biographical Memoirs of the National Academy of Sciences* 19 (1937): 51–119, on p. 56.

112. See Edward J. Larson, *Summer for the Gods: The Scopes Trial and America's Continuing Debate over Science and Religion* (Cambridge, MA: Harvard University Press, 1997).

113. Henry Fairfield Osborn, *Evolution and Religion in Education: Polemics of the Fundamentalist Controversy of 1922 to 1926* (New York: Scribner's, 1926), p. 90.

114. Henry Fairfield Osborn, "Evolution and Religion," *New York Times*, 5 Mar. 1922.

115. Quoted in Gregory, "Biographical Memoir," p. 86.

116. Gregory, "Biographical Memoir," p. 87.

117. Rainger, *An Agenda for Antiquity*, p. 42.

118. See David N. Livingstone and Mark A. Noll, "B. B. Warfield (1851–1921): A Biblical Inerrantist as Evolutionist," *Isis* 91 (2000): 283–304. See also Gary S. Smith, "Calvinists and Evolution, 1870–1920," *Journal of Presbyterian History* 61 (1983): 335–352.

119. Warfield, "Personal Recollections," p. 652.

120. B. B. Warfield, "Calvin's Doctrine of the Creation," *Princeton Theological Review* 13 (1915): 190–255, reprinted in *The Works of Benjamin B. Warfield*, 10 vols. (New York: Oxford University Press, 1927–1932), 5:287–349; citations from pp. 304–305.

121. In a critical attack on the interpretations of the author and Mark Noll, Zaspel is unwilling to take these and other comparable statements from Warfield as any indication whatsoever of Warfield's own views on evolution. See Fred G. Zaspel, "B. B. Warfield on Creation and Evolution," *Themelios* 35 (2010): 198–211. Zaspel is the author of the celebratory survey *The Theology of B. B. Warfield: A Systematic Summary* (Wheaton: Crossway, 2010). In prosecuting with such forensic tenacity the claim that Warfield allowed for evolution but did not adopt it, it may be relevant to note Zaspel's personal conviction that "the theistic evolutionist's desire to hold on to evolution and to the Bible is a groundless dream." See his "Theistic Evolution?" available at www.biblicalstudies.com/bstudy/theopropr/evlutin1.htm.

122. Bowler, *Eclipse of Darwinism*.

123. Duns's anti-Darwinism is discussed in chapter 2.

124. John Duns, review of *The Origin of Species, North British Review* 32 (1860): 455–486, on pp. 472–473.

125. Bert Theunissen, "Darwin and His Pigeons: The Analogy between Artificial and Natural Selection Revisited," *Journal of the History of Biology* 45 (2011): 179–212.

126. Warfield, *Theory and Practice of Cattle-Breeding*, pp. 86, 87, 88, 91, 92, 95, 98, 141–42.

127. William Warfield, review of "Mr. J. H. Sanders and 'The General Principles of Breeding,'" *Breeders' Gazette* (June 1885), in B. B. Warfield's scrapbook "Short-Horn Culture," Speer Library, Princeton Theological Seminary.

128. B. B. Warfield, review of *Darwinism Today* by Vernon L. Kellogg, *Princeton Theological Review* 6 (1908): 640–650.

129. See Nicolas Rasmussen, "The Decline of Recapitulationism in Early Twentieth-Century Biology: Disciplinary Conflict and Consensus on the Battleground of Theory," *Journal of the History of Biology* 24 (1991): 51–89; S. J. Gould, *Ontogeny and Phylogeny* (Cambridge, MA: Belknap Press, 1977). See also Stephen G. Alter, "The Advantages of Obscurity: Charles Darwin's Negative Inference from the Histories of Domestic Breeds," *Annals of Science* 64 (2007): 235–250.

130. B. B. Warfield, review of *The Religious Aspect of Evolution*, by James McCosh, *Presbyterian Review* 9, no. 5 (1888): 10–11.

131. B. B. Warfield, "Lectures on Anthropology" (Dec. 1888), Speer Library, Princeton University. "Evolution or Development" is reprinted in Mark A. Noll and David N. Livingstone, eds., *Evolution, Science, and Scripture: Selected Writings by B. B. Warfield* (Grand Rapids, MI: Baker, 2000), on pp. 130–131.

132. B. B Warfield, "Charles Darwin's Religious Life: A Sketch in Spiritual Biography," *Presbyterian Review* 9 (1888): 569–601, on pp. 575, 583, 590.

133. Warfield's reviews of these works are reprinted in Noll and Livingstone, *Evolution, Science, and Scripture*. Quotation on p. 152.

134. B. B. Warfield, "The Present-Day Conception of Evolution" (1895), reprinted in Noll and Livingstone, *Evolution, Science, and Scripture*, pp. 157–169, on p. 169.

135. "What is Animal Life?" *Presbyterian and Reformed Review* 1 (1890): 441–461. The contributors were John William Dawson, W. G. T. Shedd, William Berryman Scott, John Dewey, and John De Witt.

136. Gundlach, "Evolution Question at Princeton," pp. 292–293.

137. Warfield, "Evolution or Development," p. 126.

138. B. B. Warfield, "Creation, Evolution, and Mediate Creation," *Bible Student* 4 (1901): 1–8, reprinted in Noll and Livingstone, *Evolution, Science, and Scripture*, pp. 197–210, on pp. 207, 210.

139. B. B. Warfield, "The Manner and Time of Man's Origin," *Bible Student* 8 (1903): 241–252, reprinted in Noll and Livingstone, *Evolution, Science, and Scripture*, pp. 211–229, on pp. 214, 215.

140. Warfield, "Calvin's Doctrine of Creation," p. 304.

141. This is George M. Marsden's designation. See Marsden, *Fundamentalism and American Culture* (New York: Oxford University Press, 1980), p. 98.

142. Archibald A. Hodge, introduction to Joseph van Dyke, *Theism and Evolution: An Examination of Modern Speculative Theories as Related to Theistic Conceptions of the Universe* (New York: Armstrong, 1886), pp. xv–xxii, on pp. pp. xvii, xviii, xxi, xviii.

143. Archibald A. Hodge, *Outlines of Theology*, rewritten and enlarged (New York: Robert Carter, 1878), p. 39.

144. Archibald A. Hodge, review of *Natural Science and Religion*, by Asa Gray, *Presbyterian Review* 1 (1880): 586–589, on pp. 586, 587.

145. Archibald A. Hodge, review of *Natural Law in the Spiritual World*, by Henry Drummond, *Presbyterian Review* 4 (1883): 872–873, on p. 873.

146. Archibald A. Hodge and Benjamin B. Warfield, "Inspiration," *Presbyterian Review* 2 (1881): 225–260. See the discussion in Mark A. Noll, *Between Faith and Criticism: Evangelicals, Scholarship, and the Bible in America* (New York: Harper and Row, 1987).

147. Gundlach has brought this correspondence to light. See Gundlach, "Evolution Question at Princeton," p. 197.

148. Gundlach outlines the background to this chair, arguing that Patton's critique of Charles Woodruff Shields, Professor of the Harmony of Science and Religion at Princeton College, "pleased the Princeton theologians so much that the seminary created a new professorship expressly for him." Gundlach, "Evolution Question at Princeton," p. 225.

149. Francis L. Patton, "Rationalism in the Free Church of Scotland," *Princeton Review* 56 (1880): 105–124, on pp. 106, 112, 114, 115.

150. F. L. Patton, "Evolution and Apologetics," *Presbyterian Review* 6 (1885): 138–144, on p. 140.

151. Caspar Wistar Hodge Jr., Manuscript Notes on Anthropology, housed in Princeton Theological Seminary Archives. Transcribed by Bradley J. Gundlach, 1993.

152. While I have emphasized the *diversity* of Calvinist responses to Darwinian evolu-

tion, in the light of Princeton's judgment, there is much to be said for Moore's claim that Calvinist theology provided some of the best *resources* for coming to terms with Darwin. Moore, *The Post-Darwinian Controversies.*

Chapter 7 · Darwinian Engagements

1. Janet Browne, *Charles Darwin: The Power of Place* (London: Jonathan Cape, 2002), pp. 9–10.

2. Charles Darwin, *On the Origin of Species by Means of Natural Selection* (London: Murray, 1859), pp. 119, 177–178.

3. Darwin, *Origin of Species*, p. 301.

4. Gillian Beer, "Darwin's Reading and the Fictions of Development," in David Kohn, ed., *The Darwinian Heritage* (Princeton: Princeton University Press, 1985), p. 548.

5. See chapters 3 and 4.

6. Robert L. Dabney, *The Sensualistic Philosophy of the Nineteenth Century Considered* (New York: Anson D. F. Randolph, 1876), p. 187.

7. B. B Warfield, "Charles Darwin's Religious Life: A Sketch in Spiritual Biography," *Presbyterian Review* 9 (1888): 569–601, on p. 599.

8. I have discussed these in a wider context in David N. Livingstone, "Which Science? Whose Religion?," in John Hedley Brooke and Ronald L. Numbers, eds., *Science and Religion Around the World* (New York: Oxford University Press, 2011), pp. 278–296.

9. Peter Galison uses the term in his explorations of science and technology to explain how trading can take place even when the partners "ascribe utterly different significance to the objects being exchanged." Peter Galison, *Image and Logic: A Material Culture of Microphysics* (Chicago: University of Chicago Press, 1997), p. 783.

10. This draws on David N. Livingstone, "Science Wars," in Nuala Johnson, Richard H. Schein, and Jamie Winders, eds., *The Wiley Companion to Cultural Geography* (London: Wiley, 2013), pp. 371–384.

11. The other keynote speaker was to be Niles Eldredge. Keith Bennett is a colleague and informs me that Eldredge was unable to attend.

12. Keith D. Bennett, "The Chaos Theory of Evolution," *New Scientist* 208, no. 2782 (2010): 28–31, on pp. 29, 30.

13. Jerry Coyne. "Can *New Scientist* get any worse on evolution?" http://whyevolution istrue.wordpress.com/2010/11/05/can-new-scientist-get-any-worse-on-evolution/.

14. The editor's remarks can be accessed at http://whyevolutionistrue.wordpress .com/2010/11/09/new-scientist-defends-bad-science/.

15. Keith D. Bennett, "Continuing the Debate on the Role of Quaternary Environmental Change for Macroevolution," *Philosophical Transactions of the Royal Society of London*, series B 359 (2004): 295–303.

16. Jerry Fodor, "Against Darwinism," *Mind and Language* 23 (2008): 1–24, on p. 23.

17. Douglas J. Futuyma, "Two Critics Without a Clue," *Science* 328 (2010): 692–693; Massimo Pigliucci, "A Misguided Attack on Evolution," *Nature* 464 (2010): 353–354, on p. 354; Jerry Coyne, "The Improbability Pump," *Nation*, 10 May 2010; Neil Spurway, review of *What Darwin Got Wrong. ESSSAT—News* 20, no. 4 (Dec. 2010): 7–12, on p. 12; Ned Block and Philip Kitcher, "Misunderstanding Darwin," *Boston Review*, Mar./Apr. 2010.

18. Jerry Fodor and Massimo Piattelli-Palmarini, *What Darwin Got Wrong* (New York: Picador, 2010), p. xv.

19. Daniel Dennett, "Darwin's 'Strange Inversion of Reasoning,'" *Proceedings of the National Academy of Sciences* 106, supplement 1 (2009): 10061–10065, on p. 10062.

20. Robert J. Richards, "Darwin Tried and True," *American Scientist* 98, no. 3 (May–June, 2010). Available at www.americanscientist.org/bookshelf/pub/darwin-tried-and-true.

21. Michael Ruse, "Origin of the Specious," Boston Globe, 14 Feb. 2010. Available at http://articles.boston.com/2010-02-14/ae/29329727_1_natural-selection-darwinism -evolutionary-theory.

22. Fodor and Piattelli-Palmarini, *What Darwin Got Wrong*, p. xv.